SOCIAL SCIENCE AND SUSTAINABILITY

Editors: Heinz Schandl and Iain Walker

© CSIRO 2017

All rights reserved. Except under the conditions described in the *Australian Copyright Act 1968* and subsequent amendments, no part of this publication may be reproduced, stored in a retrieval system or transmitted in any form or by any means, electronic, mechanical, photocopying, recording, duplicating or otherwise, without the prior permission of the copyright owner. Contact CSIRO Publishing for all permission requests.

The moral rights of the author(s) have been asserted.

National Library of Australia Cataloguing-in-Publication entry

> Social science and sustainability / Heinz Schandl and Iain Walker (editors).

> 9781486306404 (paperback)
> 9781486306411 (epdf)
> 9781486306428 (epub)

> Includes index.

> Social sciences.
> Sustainability.

> Schandl, Heinz, editor.
> Walker, Iain, editor.

Published by

CSIRO Publishing
36 Gardiner Road, Clayton VIC 3168
Private Bag 10, Clayton South VIC 3169
Australia

Telephone: [+613] 9545 8555
Email: csiropublishing@csiro.au
Website: www.publishing.csiro.au

Front cover: Solar panel (credit: Alan Levine/Flickr, CC BY 2.0); people crossing street (credit: Mark Gunn/Flickr, CC BY 2.0)

Set in 10.5/12 Minion Pro & ITC Stone Sans Std
Edited by Adrienne de Kretser, Righting Writing
Cover design by Andrew Weatherill
Typeset by Desktop Concepts Pty Ltd, Melbourne
Index by Indexicana
Printed by Ingram Lightning Source

CSIRO Publishing publishes and distributes scientific, technical and health science books and journals from Australia to a worldwide audience and conducts these activities autonomously from the research activities of the Commonwealth Scientific and Industrial Research Organisation (CSIRO). The views expressed in this publication are those of the author(s) and do not necessarily represent those of, and should not be attributed to, the publisher or CSIRO. The copyright owner shall not be liable for technical or other errors or omissions contained herein. The reader/user accepts all risks and responsibility for losses, damages, costs and other consequences resulting directly or indirectly from using this information.

Feb26_RP_ILS

Foreword

'Sustainability' is inherently an integrative concept. It encourages us to link together different dimensions of value (economic, environmental, social) and take action to sustain them over time.

The biophysical sciences are an obvious partner of sustainability. Typically, we turn to scientific inquiry to understand the consequences of our actions on our planet as we consume its resources to support our development. Science helps us to innovate so we can deliver maximum value and minimum impact from this consumption. Science, then, can provide us with the knowledge to deliver 'sustainability', if we can work out what we wish to sustain. The trouble is that the definition of value and the relative importance of economic development, environmental preservation and social welfare vary for different people and at different scales. Working out what we wish to sustain and taking action to do so is a social issue.

The social sciences and humanities give us a rich tapestry of knowledge, experience and tools to understand the culture and history of our societies, how they work, how we relate to each other and what matters to us and to others. It is the social sciences that provide us with the tools to understand what is meant by 'value' and who wins and who loses from different courses of action. The chapters in this book are written by researchers expert in applying the social sciences to determine just this. And it is not straightforward. This is a highly complex field of study that has evolved dramatically over the past decade.

Like many trained in the biophysical sciences, I started my journey into the field of sustainability wishing that the social sciences would develop its own quantitative and measureable frameworks that could stand proudly alongside economic evaluations of net present value, or environmental indicator systems such as water quality. In this manner, we could simultaneously evaluate economic, environmental and social impacts and determine the course of action that maximised value for the majority. Having had the privilege of working alongside many of the authors in this book, I now appreciate the naivety of that view, which completely underestimates both the significance of the social sciences to the concept of sustainability and the enormous epistemological differences between the economic, social and environmental sciences.

The chapters that follow highlight this. Presented through a range of case studies from different sectors, the articles draw on CSIRO's long history in sustainability science and present a modern take on the role of the social sciences in delivering sustainability. The chapters focus on some of the critical social issues that will fundamentally determine society's movement (or not) towards sustainability – addressing the diversity of societal needs and values in determining appropriate action towards sustainability, building social capacity to adapt in the interests of sustainability and building the institutions and behaviours to drive action towards sustainability.

Amory Lovins identifies three distinct foci for power and action in the world: business, civil society and government. These are all social constructs which have emerged from the historical and cultural evolution of societies over 4000 years. Out of the social sciences, the concept of rational economics emerged. This defines value in monetary terms around goods and assets, has driven the development of western civilisation and drives the performance of our most globally influential present-day institutions. And it is out of the social sciences that frameworks will emerge that broaden our definition of value and help us move toward new paradigms promoting sustainability.

So while the biophysical sciences enable us to make sense of observations and phenomena in the physical world, it is the social sciences that will help society organise to sustain important values in a changing world. The social sciences do not sit comfortably alongside other sciences and that is precisely why they are so important for sustainability. They enable us to make sense of what we observe in human relationships, between individuals, groups, nations, sects, interest groups and political institutions at different scales and over time. Social science should not be considered a minor partner in the drive for sustainability. It is the discipline that will determine whether we succeed or fail in our attempts.

Anna Littleboy
Research Director | Resources, Community and Environment
CSIRO Mineral Resources
January 2017

Contents

Foreword		iii
Acknowledgements		vii
Contributors		viii

1 Introduction — 1
Heinz Schandl and Iain Walker

2 Why do we need a sociology of society's natural relations to inform sustainable development? — 9
Heinz Schandl

3 Integration science for impact: fostering transformations towards sustainability — 23
Ro Hill, Cathy Robinson, Petina Pert, Marcus Barber, Ilisapeci Lyons, Kirsten Maclean, Leah Talbot and Catherine Moran

4 Integrating development studies and social-ecological systems thinking: towards livelihood adaptation pathways — 51
James R.A. Butler, Liana J. Williams, Toni Darbas, Tanya Jakimow, Kirsten Maclean and Clemens Grünbühel

5 Remote, marginal and sustainable? The key role of brokers and bridging institutions for stronger Indigenous livelihoods in Australia's deserts — 75
Jocelyn Davies, Yiheyis T. Maru, Fiona Walsh and Josie Douglas

6 Sustainability science, place and regional differences: vulnerability and adaptive capacity in Sydney — 99
Tom Measham, Bruce Taylor and David Fleming

7 A hierarchy of needs for achieving impact in international Research for Development — 109
James R.A. Butler, Toni Darbas, Jane Addison, Erin L. Bohensky, Lucy Carter, Michaela Cosijn, Yiheyis T. Maru, Samantha Stone-Jovicich, Liana J. Williams and Luis C. Rodriguez

8 **The co-construction of environmental (instream) flows and associated cultural ecosystem benefits** — 131
Rosalind H. Bark, Cathy J. Robinson, Sue E. Jackson and Karl W. Flessa

9 **Dipping in the well: how behaviours and attitudes influence urban water security** — 145
Anneliese Spinks, Kelly Fielding, Aditi Mankad, Rosemary Leonard, Zoe Leviston and John Gardner

10 **Making sense of Australians' responses to climate change: insights from a series of five national surveys** — 161
Iain Walker, Zoe Leviston, Rod McCrea, Jennifer Price and Murni Greenhill

11 **Innovation, sustainability and the promise of inclusion** — 177
Lucy Carter

12 **Risk, sustainability and time: sociological perspectives** — 187
Stewart Lockie and Catherine Mei Ling Wong

13 **Policy-relevant research: improving the value and impact of the social sciences** — 199
Brian W. Head

Index — 211

Acknowledgements

This book celebrates the important scientific work done on sustainability by members of the Social and Economic Sciences (SES) Program of Australia's Commonwealth Scientific and Industrial Research Organisation (CSIRO). Some of that work appears in this book; much of it lies behind the scenes. It is right and proper for us as editors of this book to acknowledge, respect and admire the splendid, cutting-edge contributions our colleagues have made over a long period. We have been fortunate to have had so many fine colleagues over the years. Marcus Lane deserves special mention here. He was instrumental in developing and running the SES Program to become perhaps the world's largest collection of social scientists working on sustainability. His relentless focus on doing high-quality social science in this space drove a step-change in the quality of social science work being produced by CSIRO. Much of the work presented in this book is a legacy of Marcus' energy, foresight and emphasis on science quality with impact.

The book was preceded by a two-day symposium held in Brisbane in May 2014. Kristi-Ann Fenech and Margot Simpson were pivotal to organising and running the symposium smoothly and efficiently. Dan Walker, as Chief of the Division of Ecosystem Sciences in which the SES Program operated, supported the symposium, as he has supported the social sciences generally.

Turning a symposium into a book is no easy task. Each of the chapters has benefited from the reviews provided anonymously through CSIRO's internal review process. Lauren Webb of CSIRO Publishing has supported the idea of this book and nurtured it into being. Editors have to be patient people; Lauren must be one of the most patient of all. Craig James, the current Research Director of the Adaptive Urban and Social Systems Program, the latest organisational incarnation of the SES Program, has provided support for Karin Hosking to work with us as project manager and technical editor. We have indeed been fortunate to have Karin's support, skill and patience at all stages of the production of this book.

Contributors

Jane Addison is a scientist of natural resource management, with a particular interest in ecological variability and institutional design. Jane is currently lecturing at James Cook University, and has ongoing involvement in a research for development project in Mongolia and China that seeks to improve landscape condition and pastoral livelihoods. Previously, she was with the CSIRO Townsville and Alice Springs.

Marcus Barber is an environmental anthropologist. His key research interests include the interface between scientific and indigenous knowledges, water issues, sustainable development and conservation-based livelihood options for Indigenous Australians. He is an Adjunct Senior Lecturer in the School of Social Science at the University of Queensland.

Rosalind H. Bark is a European Commission Marie-Curie Fellow at the University of Leeds. Previously she was a senior research scientist at CSIRO. She completed a PhD and post-doctoral research at the University of Arizona. Her research interests are river basin water resources management in contested environments, valuing water-dependent ecosystems, tracking socio-cultural aspects, and operationalising the ecosystem approach.

Erin L. Bohensky, PhD, is a sustainability scientist who conducts research on engagement with and adaptation to social and environmental change. Her work explores the theory and practice of knowledge integration and co-production through participatory processes, and has assisted governments and communities in Australia, Papua New Guinea, Indonesia and Africa to understand and respond to complex social-ecological challenges.

James R.A. Butler is a sustainability scientist trained in agricultural economics, terrestrial and marine ecology in southern Africa, Europe and Australia. He analyses complex development problems in the Asia-Pacific region, focusing on trans-boundary issues linking northern Australia, Indonesia and Papua New Guinea. He applies concepts of resilience, transformation and well-being to explore alternative livelihood development pathways through participatory action research.

Lucy Carter is a philosopher and social scientist with a background in international health, agriculture–nutrition links, and the ethics of development research and practice. She has held roles in research and management across multiple sectors including biotechnology, public policy and health. She holds a PhD from the University of Queensland and additional post-graduate qualifications in health from Curtin University and the University of Queensland.

Michaela Cosijn holds an MSc in integrated Environmental Sciences from Southampton University and is a research scientist at CSIRO. Her key skills are in climate adaptation and proofing of programs, pro-poor value chain development in the agriculture sector, agricultural innovation systems, establishing monitoring and evaluation systems, participatory planning processes, and partnership brokering. She has worked in sub-Saharan Africa, West Africa, Asia, Australia and New Zealand.

Toni Darbas is a UNSW-trained political sociologist who has worked for the past 12 years within interdisciplinary sustainability science with hydrologists, agronomists, ecologists and others. Her work is focused at the interface between local communities and formal extension systems that is problematised by policy efforts to achieve natural resource management, rural development and agricultural innovation in Australia, South-east and South Asia.

Jocelyn Davies is a geographer whose work has strengthened recognition of Indigenous land rights and land management, particularly in Australia's arid rangelands and protected areas. By facilitating disciplinary integration, she has contributed to CSIRO's efforts to overcome constraints on food security in sub-Saharan Africa. After a decade as principal research scientist with CSIRO, she is a University Fellow with the Northern Institute, Charles Darwin University, Alice Springs.

Josie Douglas, PhD, is descended from the Wardaman people in the Northern Territory and works as a Senior Policy Officer with the Central Land Council in Alice Springs. Josie was awarded a doctorate in 2015 for her research, conducted while she was with CSIRO, examining the intergenerational transmission and acquisition of Indigenous ecological knowledge in contemporary contexts in central Australia.

Kelly Fielding is an Associate Professor at the University of Queensland. Her research focuses broadly on understanding the social and psychological determinants of environmental sustainability. She seeks to understand environmental decisions and behaviours and to develop communication and behaviour change strategies that can promote greater environmental sustainability.

David Fleming, PhD, is an economist with interests in economic growth, regional development and applied econometrics. In Australia, he has provided important insights into the economic impacts of coal seam gas mining in Queensland and on the benefits (and costs) of the recent mining boom across the country. He has participated on the editorial board of prestigious journals and has published in leading journals such as *Food Policy* and the *Journal of Development Studies*.

Karl W. Flessa is a Professor of Geosciences and conservation biologist at the University of Arizona in Tucson. He works with a team of Mexican and US biologists, hydrologists and remote-sensing specialists to examine the effects of the first major transboundary environmental flow to the Colorado River delta. He was a CSIRO Distinguished Visiting Scientist in Brisbane in 2013–14.

John Gardner has a background in social psychology and research consultancy. He works in the domains of energy, water and climate adaptation, investigating the factors and interventions that motivate individuals, communities and organisations to change consumption, adopt new technologies and approaches, and reduce environmental impacts.

Murni Greenhill is a social scientist working with CSIRO Land and Water in Perth, Western Australia. She has an academic background in psychology. Her research interest is in understanding how people's decision-making processes are shaped by their belief and value systems, and what impacts these have on the natural environment.

Clemens M. Grünbühel, PhD, is an ecological anthropologist with expertise in sustainable resource use and development. His research interests include climate adaptation, rural innovation, community development and resource use systems. He is a faculty member of the School of Environment, Resources and Development, Asian Institute of Technology in Bangkok.

Brian W. Head has a PhD in political science and is ARC Professorial Fellow at the University of Queensland. He has held senior roles in government, academia and the non-government sector. His work with the Queensland government included environment policy, public sector reform and intergovernmental programs. His major interests are evidence-based policy, program evaluation, complex policy challenges, collaboration, accountability and leadership.

Ro Hill is a human geographer dedicated to collaborative environmental governance, multiple evidence approaches, Indigenous peoples and sustainability with over 50 relevant peer-reviewed publications. She serves on the Intergovernmental Platform on Biodiversity and Ecosystem Services Expert Taskforce for Indigenous and Local Knowledge and is a Research Team Leader with CSIRO Land and Water.

Sue E. Jackson is a geographer with over 20 years' experience researching the social dimensions of natural resource management, particularly community-based conservation initiatives and institutions. She has research interests in systems of resource governance, including customary Indigenous resource rights and Indigenous capacity-building for improved participation in natural resource management and planning, as well as the social and cultural values associated with water. She is an Associate Professor and ARC Future Fellow based at the Australian Rivers Institute, Griffith University.

Tanya Jakimow is a Senior Lecturer of Development Studies and ARC Research Fellow (DECRA) at the University of New South Wales. Her latest book, *Decentring Development: Understanding Change in Agrarian Societies* (Palgrave Macmillan, 2015) examines agrarian change in Telangana, India and Central Lombok, Indonesia.

Rosemary Leonard is a Professor at the University of Western Sydney where she holds a UWS–CSIRO Joint Chair in Social Capital and Sustainability. She completed her doctoral work at Macquarie University in social psychology. Since joining CSIRO, she has led the social component of a number of research projects examining people's understanding of, attitudes to and engagement with alternative water systems.

Zoe Leviston is a research scientist at CSIRO in Perth. Her research investigates how individuals, groups and culture shape people's responses to natural resource challenges. Her application of social psychology across environmental domains seeks to reveal how elements such as social identity, personality traits and beliefs about social behaviour influence how we process information about the environment.

Stewart Lockie is Professor of Sociology and Director of the Cairns Institute at James Cook University. His research addresses environmental policy, food and agriculture, social impact assessment, poverty alleviation and emerging issues facing the global sustainable development agenda.

Ilisapeci Lyons, PhD, is a post-doctoral social scientist with CSIRO. Her research title is Scientific and Indigenous knowledge integration: testing conditions for effective biodiversity management. Prior to joining CSIRO, she worked with the Queensland government on natural resource management programs and undertook research in the Pacific and South-east Asia on community-level governance and community engagement and development.

Kirsten Maclean is a human geographer with the CSIRO Adaptive Urban and Social Systems program. She uses co-research practice and participatory methodologies to investigate the role of diverse people, knowledge and values in informing regional natural and cultural resource planning and management in Australia.

Aditi Mankad is an interdisciplinary research scientist with CSIRO Land and Water. Her academic training is in sport, exercise and health psychology. Aditi's research utilises social psychological theories of motivations, cognitions and behaviours to help understand real-world challenges associated with alternative resource use and agricultural practices, as well as acceptance of novel technologies and innovations within these domains.

Yiheyis T. Maru is a social-ecological systems scientist with a Doctor of Veterinary Medicine degree and a PhD in systems thinking research and practice. He has extensive skills and experience in interdisciplinary research for food security, biosecurity, sustainable livelihood development and enabling adaptation and transformation pathways in developing countries and in Australia, especially among indigenous peoples.

Rod McCrea is an interdisciplinary social scientist with CSIRO Land and Water in Brisbane. With an academic background in psychology, economics and human geography, his research interests are at the interface of environment and behaviour, focusing on environmental psychology, community well-being, and adaptability in the context of resource and environmental sustainability issues.

Tom Measham is a human geographer who has worked extensively in interdisciplinary research teams with sociologists, economists and natural scientists (among others) on the ways in which place-based communities engage with global challenges and resource management. He has a PhD from the Australian National University, an MSc from James Cook University and a BA from the University of Sydney.

Catherine Moran is an ecologist and knowledge-broker with research interests in the impacts of global change on human and natural communities, as well as opportunities for adaptation and resilience.

Petina Pert has worked extensively in interdisciplinary research teams with sociologists, economists and ecologists. Her work informs natural resource management decisions utilising her GIS and remote-sensing expertise and spatial analysis skills. She has a PhD from RMIT University and a Master of Applied Science (Parks, Recreation and Heritage Management) from Charles Sturt University.

Jennifer Price has 10 years' experience in applied research on environmental issues, specialising in psychometrics and theoretical development. Her research focuses on human–environment interactions, revealing the role of cultural values and time orientation in decision-making. This involves social-psychological research into climate change mitigation and adaptation, agricultural practices and water treatment. Her work shows how identity, personality and context intersect to influence risk responses.

Cathy J. Robinson, PhD, is a human geographer specialising in the design of governance systems that enable different types of knowledge to be useful for natural resource policy and decision-making. Cathy is a Principal Research Scientist at CSIRO, Professor and Deputy Director of the Northern Research Institute at Charles Darwin University, and IUCN Commissioner for the World Commission on Protected Areas.

Luis C. Rodriguez has a PhD in ecological economics and was a senior researcher for CSIRO in Brisbane. He is currently an advisor for the performance evaluation of the Forest/Environment components of the Free Trade Agreement Peru–USA, a member of the scientific and technical committee of the Inter-American Development Bank for the impact evaluation of poverty-environment initiatives in Ecuador, and a member of the Global Expert Group on Integrated Environmental Assessment for UNEP.

Heinz Schandl has a PhD in sociology. He is a senior scientist at CSIRO in Canberra, honorary professor at the ANU Fenner School of Environment and Society and adjunct professor at Nagoya University in Japan. He is a member of the UNEP International Resource Panel. His research connects social theory, social metabolism and environment and sustainability policy.

Anneliese Spinks is a social scientist with research expertise in the areas of epidemiology, environmental psychology and the social determinants of health. She also contributes to teaching and research integrity through involvement in human research ethics and voluntary teaching of public health modules to students from low- and middle-income countries.

Samantha Stone-Jovicich has a background in environmental sociology and social-cultural anthropology. Her research in CSIRO focuses on understanding how non-technical dynamic approaches (such as cross-scale partnerships, social learning processes and innovation systems approaches) to addressing global social-environmental challenges can be designed and implemented for greater impacts. Her primary interest is in systemic change and the role science and research can play in contributing to wide-scale and lasting social-ecological changes.

Leah Talbot is with Land and Water, CSIRO, currently completing her PhD in the integration of conservation and Indigenous knowledge, governance systems, rights and interests. As an Indigenous person of far north Queensland, her experiences include conservation and environmental management, high-level Indigenous negotiations and developing collaborative Indigenous research methodologies and participative planning with Indigenous communities.

Bruce Taylor is a human geographer with research interests in environmental policy implementation and governance in both rural and urban settings. His work informs regional and national planning for sustainable development, cooperative approaches to natural resource management, and adaptation to climate change in coastal and urban Australia.

Iain Walker is a social psychologist and Professor of Psychology at the University of Canberra. He was formerly a senior principal research scientist at CSIRO and Research Program Leader of the Social and Economic Sciences Program.

Fiona Walsh is an ethno-ecologist with a PhD in botany and anthropology. For 28 years she has worked among Indigenous Australian groups to collaborate, lead or contribute to cross-cultural programs. As a co-researcher and community-focused facilitator, she has strengthened the confidence and capabilities of Indigenous individuals and groups and shaped the direction of major cultural heritage and land management projects. She is based in Alice Springs and is a former CSIRO research scientist.

Liana J. Williams is a human geographer concerned with social and institutional implications of agricultural and rural development in South-east Asia. While working at CSIRO she is also undertaking a PhD at the University of Queensland, concerned with the processes and impacts of applying findings from agricultural research projects at scale.

Catherine Mei Ling Wong is an environmental sociologist working on risk governance in the climate, energy and financial sectors. She did her first post-doc at The Cairns Institute on climate governance and currently works on financial governance in global cities and sustainable urban development at the University of Luxembourg. She has a PhD in sociology from the Australian National University, and a Master in Global Studies from the University of Freiburg, Germany.

1

Introduction

Heinz Schandl and Iain Walker

Many would agree that the global environmental crisis has reached new and concerning levels with climate change spiralling out of control, and other environmental issues of natural resource depletion, pollution and growing waste converging rapidly, leading to an unsustainable global state. It seems that globally thresholds are being passed with little hope of return. At the same time, billions of people aspire to a better standard of living, and rightly so. One wonders how this will be achievable with a natural resource base that is already stretched, and environmental integrity that is severely compromised. Harmonising economic and social development goals and environmental goals is more important than ever.

In response to the dual objectives of development and environment, the international policy community has just renewed its commitment to sustainable development, expressed in the 17 new Sustainable Development Goals that aim to achieve global human well-being for all, within the limits of global boundaries. When we compare the new goals to the previous Millennium Development Goals, it is striking how strongly they refer to environment and natural resources as a priority. Sustainable management of natural resources and well-functioning ecosystems are understood as fundamental to achieving any of the human development outcomes countries have signed up for.

There is also international consensus, achieved at the Paris Climate Change conference in 2015, that the global community needs to invest more to combat climate change and to ensure greenhouse gas abatement policies are sufficient to deliver a world below 2°C of warming. It will require a massive effort to decarbonise the global energy-generation infrastructure. Yet, despite all the high-level commitments and good intentions, change is occurring only slowly, and individual aspirations and everyday life are not aligned with the big policy strategies.

The quest for sustainability occurs against the backdrop of the largest and fastest industrial and urban transition of the developing part of the world ever experienced in human history. By 2030, the world expects 3 billion additional middle-class consumers. The second wave of urbanisation in Asia, Africa and Latin America will require an 80% rise in steel and cement production by 2030 and the cost of extraction of oil and metals will double. This will mean that the world market price for energy, metals and food will increase again, and prices for primary materials may reach the levels of 2008 at the onset of the global financial crisis. To satisfy the new demand for natural resources, large investments into energy, water and resource supply systems will be needed.

At the same time, there is considerable potential for resource and emissions savings in many systems of provision including construction and housing, transport and mobility, agriculture and food. The same is true for energy supply and heavy industry. Many services and products can be delivered with savings of up to 80% using current technology. The recycling potential of many metals and other materials is also underutilised, offering a large potential for developing a circular economy where waste flows become inputs into industrial process, thus saving virgin resources. It appears straightforward that investment would need to be redirected to achieve human well-being at much lower environmental cost.

The next wave of growth and innovation may well lie in the new green sectors of the economy, yet we often see more of the same. What are the reasons for inertia? Why are resource efficiency and climate mitigation technologies and sustainable urban design not picked up more readily? Why has everyday life not adjusted to the needs of global sustainability – an outcome which, if achieved, would have many benefits for people and the planet? How can we guide change towards sustainable consumption and production? How can society and economy be steered towards sustainability, what changes to governance are required, and who would make it happen and when? These are all core questions for the social sciences.

In this book we start from the premise that the connections and interdependencies between environmental and social and economic systems are core to all sustainability problems confronting the world today and, more importantly, are central to developing solutions to those problems. To date, though, there is no coherent social science of sustainability. The social sciences themselves are complex and disparate, often rest on different frameworks and methods, yet still have unique contributions to make to sustainability challenges.

When a new interdisciplinary 'discipline' of sustainability science was launched with a seminal contribution in the journal *Science* (Kates et al. 2001), the authors asked a set of interdisciplinary questions which remain relevant. Sustainability science starts from an analysis of coupled social and natural systems, whose interactions are shaped through the way production and consumption systems are structured and how they rely on natural resources and produce waste and emissions. It includes the analysis of vulnerabilities in the society–nature system, the need for resilience and for staying within the boundaries of well-functioning society–nature systems. Most importantly, and of great relevance to the social sciences, is the core question of sustainability science – how to guide the interaction between society and nature towards more sustainable trajectories. The authors of the seminal paper expressed the need for a science of sustainability that informs policy and business decisions to promote sustainable development at the global scale.

These questions are considerable. They require major science breakthroughs that will likely only come from investment in inter- and transdisciplinary efforts done at scale on issues of national and international significance. The Commonwealth Scientific and Industrial Research Organisation (CSIRO) is Australia's national science agency and has been a main contributor to sustainability science. Enabled by the unique way in which CSIRO can organise large research teams to engage with major scientific and policy problems, interdisciplinary teams – social scientists working alongside climate, water and ecosystem scientists – have been established. Recognising the importance of social sciences in addressing environmental sustainability has required a large intake of social scientists into an organisation historically dominated by biophysical scientists and engineers.

Over the past decade or more, CSIRO has invested significantly in developing social science contributions to sustainability, drawing on bodies of knowledge from sociology, human geography, social psychology, anthropology and economics. CSIRO's social science capability has grown quickly, from just 12 social scientists in 1985. It now has Australia's largest single grouping of social scientists working on environmental and sustainability issues. Interdisciplinary collaboration between social and natural sciences is not always a straightforward process. It requires investment in conceptual frameworks that have relevance in both the social and natural sciences, and the design of research questions that are meaningful and scientifically interesting across both domains. New knowledge generated between the disciplines needs to be reintegrated and to remain plausible within the parent scientific communities. For the social sciences, this has sometimes required the contestation of established paradigms to fully appreciate the new knowledge. Social scientists in CSIRO and elsewhere need to cease being an afterthought to the biophysical sciences, and to avoid being seen as the 'socialising sciences', brought in at the end of projects to deliver science communication or community engagement aspects of projects. This book showcases some of the work of CSIRO's social scientists and aims to bring conceptual clarity to the role of social sciences in informing sustainability analysis, policies and practices in interdisciplinary collaboration with biophysical scientists.

The contributors to this book, to different degrees, summarise large programs of sustainability research they have led or been involved in, distil generalisable practical and conceptual lessons, give overviews and critiques of the theoretical frameworks that lie behind those research programs, and address the translation of social science into policy- and decision-making.

To what extent can the experience from a research organisation which has allowed social scientists to engage with natural scientists in the environment and sustainability domain be integrated into a broader, coherent social science contribution to the field of sustainability? How can socio-economic sustainability science be made available to policy- and decision-makers around the globe to improve environmental and sustainability outcomes? These are the pertinent questions addressed in this book to inform a policy domain that is characterised by the complexity of the issues, contestation about possible solutions, incomplete knowledge about the consequences of many interventions into environmental systems, and sometimes ignorance. Complex issues require theory, however, that is appropriate for the level of complexity of the problem, and empirical research that informs new thinking about how to improve the ways in which society can steer its environmental relations to become more sustainable.

Theory and practice of social sustainability science

Human activity, individually, organisationally, institutionally and systemically, shapes and is shaped by ecosystems large and small. How human systems and ecosystems interact with each other, co-constitute each other, is the stuff of the sustainability sciences. There are several frameworks to conceptualise how to research sustainable development and the intersections and interactions between human and environmental systems. Arguably, none fully captures simultaneously the complexity of both sides of the social-natural dialectic. Chapter 2 draws on the theoretical developments of two German sociologists, Luhmann and Habermas, to describe the challenge of developing a grand theory of social-ecological transition. A major value of a sociological perspective in the sustainable development agenda is that it shifts focus away from individual actions and decisions, presumably motivated to maximise individual utility, and places it squarely on socio-technical systems, how such systems enable or prohibit choices and behaviours, and how such systems and their institutional arrangements can be transitioned to more sustainable configurations.

The need to understand changes in social systems connects to Chapters 3 and 4. The focus of Chapter 3 (by Hill *et al.*) is transformation – transformation of scientific knowledge and how it is produced; transformations in how science is done, in partnership, in social contexts; and ultimately transformations in social systems themselves to produce more sustainable arrangements. Hill *et al.* summarise experiences from eight different case studies to demonstrate the utility of integration science in fomenting such transformations. Most of their case studies involve Indigenous collaborators. The chapter highlights the importance of ethics, and argues cogently for the recognition and inclusion of Indigenous knowledge systems in any effort to produce transformation towards sustainability. Marginalisation, it contends, is not just that of social groups from power and privilege, but also of knowledge and knowledge systems.

Chapter 4 (by Butler *et al.*) presents a different but related approach to conceptualising transition pathways to sustainable systems. It examines the relative adequacies of the sustainable livelihoods framework and the resilience-thinking framework, both commonly used in the international literature on research for development (and elsewhere), using a case study of a village in Thailand to explore the attributes of each framework. Despite significant commonalities between the two frameworks, their slight differences in relative focus – on people or on community and governance, respectively – suggest different interventions to bolster sustainable practices. It concludes with some speculations about the benefits and limitations of using a hybrid model in participatory decision-making contexts.

The sustainable livelihoods framework is also the focus of Chapter 5 (by Davies *et al.*) but in a rather different way. The chapter analyses Indigenous livelihood activities in central Australia, using the framework and the 'five capitals' approach, to highlight the essential role played by 'brokers', i.e. people who link across structural holes in social networks. As the chapter points out, the livelihoods framework encourages a focus on assets and social value, rather than the deficit approach implicit in so many of the policies and practices of mainstream Australia. Brokers are an asset in and for remote Indigenous communities, but themselves face institutional and other problems.

'Sustainability' is an abstract concept, perhaps easy to recognise (especially in its absence) but almost impossible to define. Chapter 6 (by Measham *et al.*) may not help us define sustainability, but it does locate it. Sustainability has to be about a place, it reasonably contends. The chapter summarises the results of a long-term collaboration with local authorities in Sydney, mapping spatial variability in local and regional social vulnerabilities to climate-change disasters. Importantly, local social characteristics affected the potential for harm from disaster as much as biophysical ones did. Places, the chapter concludes, are where social systems, cultural systems and biophysical systems come together.

An important feature of the project described in Chapter 6 was the collaboration between scientists and local authorities, with scientists iteratively shifting between research and those responsible for managing risk and developing adaptive capacity. Scientists here study vulnerability and change, but also help catalyse change through that work, and are themselves a resource contributing to regional adaptive capacity. The role of scientists is also a focus of Chapter 7 (by Butler *et al.*), which describes work on research for development (R4D) in developing countries. The intersection of social and environmental systems in developing countries (and elsewhere) is a complex system. To pursue and deliver R4D in these contexts, scientists have to change their traditional science practices, reflexively contemplate their own role as part of the complex system being examined, and consider the process of local engagement as much as the content of their science. Overviewing and synthesising their work and reflecting on their experiences in seven different R4D projects, the chapter proposes a seven-step 'hierarchy of needs' to ensure that research can deliver impact and value to local communities in developing countries. Following this hierarchy provides a powerful, but intensive and expensive, way of conducting research with impact on people's livelihoods in R4D contexts.

Complexity of a different sort is a hallmark of the management of major river basins around the world, especially across state and national borders. The importance to water management and governance of social, economic and cultural values associated with river systems is the focus of Chapter 8 (by Bark and colleagues). It uses a case study of the 2014 pulse flow along the Colorado River, which was the first time the river had reached the sea since 1998, and an analysis of the media coverage of the flow. The analysis showed that cultural processes are grounded in the interactions between humans and their environment, that ecosystem services have a socio-cultural basis, and that the pulse flow heightened people's experiences of the cultural values associated with the river.

Water is also the focus of Chapter 9 (by Spinks *et al.*) but in a very different way. Australia is a dry continent, and water security is a perennial concern. The chapter reviews six major projects on water and water supply in Australia, spanning efforts to curb domestic consumption, examining barriers to the use of alternative water sources, the acceptance of decentralised water systems, and the acceptance of water-sensitive urban design. Just as Chapter 8 highlights the importance of cultural values around water, Chapter 9 shows that the relationships between people and water, and how people behave with water, are always influenced by broader social and technical systems.

The focus on individual beliefs, attitudes and behaviours carries into Chapter 10 (by Walker *et al.*) Here, the focus is on understanding the Australian public's opinions about climate change and how those opinions relate to behaviours. Summarising highlights from a series of five annual national surveys, the chapter demonstrates that opinions about climate change are both stable and full of contradictions, that people have poor knowledge of other people's opinions, and that people's opinions are both a cause and a consequence of environmentally relevant behaviours, including voting. To make sense of these results, the chapter draws upon a framework that sees opinions about climate change as expressions of one aspect of how people understand their relationship with the social and environmental worlds around them. Opinions are functional, psychologically and socially; they are systems in tension and change.

Chapters 9 and 10 both draw heavily on psychology and focus largely on individual thoughts and behaviours, albeit in groups and in context. Individuals are important agents, to be sure, but they are not disconnected atoms – individuals occupy roles, either assigned or attained, and are enabled and constrained by the social and organisational positions they hold. Chapter 11 (by Carter) offers a personal reflection on the author's experiences pursuing innovative approaches to sustainability science in large projects on food systems research in developing countries, with particular focus on gender parity and inclusion. The project revealed the essential role of the broker in making innovative research on food systems work. This role, along with a range of activities mostly unfamiliar to scientists but necessary for the success of projects like this, presents opportunities, costs and opportunity costs for anyone filling the role. The nature of the role can be seen to be stereotypically 'feminine', and hence it may be more likely that women will come to occupy the role, either by choice or by assignment. Such work is not usually rewarded in or by science or science organisations, though, producing a 'gender penalty'. Things do not have to be that way. Doing innovative systems research in sustainability requires innovation not just in the science, but in the institutional conduct of science and other organisations involved.

The final two chapters in the book present somewhat of a shift in approach. The authors of Chapters 1–11 are current or former members of the Social and Economic Sciences (SES)

Program and its successors, presenting work done through CSIRO. Social scientists in the Program work in cross-organisational contexts, though, and their science is not isolated from the rest of the scientific community. Chapters 12 and 13, the final two chapters, take us out of the SES Program and offer something of an external reality check.

Chapter 12 (by Lockie and Wong) considers the social organisation of risk. The construction and management of risk arguably cuts across every chapter in the book so far, and indeed cuts across all the social sciences, from individual psychology to the sociology of power, governance and public policy. What constitutes a risk to a society is not a result of a probabilistic assessment of the likelihood and severity of a negative event; rather, it is a moral judgement about social order, distributed unevenly spatially, temporally and socially. The chapter calls for greater and more sophisticated analyses of risk and its management across the social sciences, and requires the development of the potential of participatory and other approaches.

Finally, Chapter 13 (by Head) considers the translation of social science research on sustainability into public policy. In some ways, the impact on public policy of the research described in this book is a litmus test of whether the research is attaining CSIRO's legislated mandate of 'furthering the interests of the Australian community'. Although scientists – even social scientists – are often poorly equipped for engaging with policy, and even naïve about how to go about influencing policy, the social sciences generally have been making broad contributions in many diverse, sometimes indirect, ways.

A coherent social science contribution?

The contributions in this book draw from a variety of social science disciplines with diverse conceptual backgrounds, methods and epistemologies. They all describe enduring programs of research engaging in problems of society–nature interaction and how these interactions can be steered towards environmentally and socially sustainable outcomes. To social scientists, the differences between the social sciences appear stark; to natural scientists, policy-makers and others, those differences appear trivial.

What unifies the social sciences when they engage with natural scientists and others in sustainability science research, and which notions from the social sciences remain viable in the interdisciplinary context?

First, the social sciences see environmental problems not as problems of individual behaviour. They prefer analyses at a higher level of social organisation, whether referred to as system, structure or institutions. This acknowledges that throughout human history the process of interacting with nature and natural resources and of transforming natural systems has been socially organised and has become more and more removed from individual decisions. Attempts to change social-natural relations, as a consequence, rely on concerted action, and social choices and interventions need to be targeted at the system and institutional make-up of our social arrangements. This is not to deny the importance of understanding individual choices and behaviours. These too can – and must – be changed if we are to move towards more sustainable forms of social life, but they can be neither understood nor changed in atomised isolation.

Second, sustainability problems are intrinsically linked to social organisation, forms of governance and modes of production and consumption. They can be expressed functionally or geographically. Both perspectives are required and need to be harmonised in some fashion to grasp the full complexity and multi-layered (across scale) characteristics of sustainability problems. The sustainability of any social arrangement is a result of cultural evolution vis-à-vis nature and a direct result of social interaction and modification of natural systems, i.e. of history. The course can be changed, perhaps, through collective effort and social choices expressed through polices that enable human well-being at much lower and less steady resource flows and emissions and lower levels of interference in the functioning of natural ecosystems.

Third, interdisciplinarity and transdisciplinarity are key to exploring the prospects of sustainability, to allow for experimentation and to challenge the status quo of unsustainable ways of production and consumption. This will include the integration of scientific knowledge with other bodies of knowledge from multiple places in society. It also requires a collaboration between the social and physical sciences which goes well beyond additive and multidisciplinary contributions and engages with new conceptual and analytical frameworks that capture the complexity of the interaction of social and natural systems. There are many examples of how this can be achieved, and they need to start influencing communities outside the science community.

Fourth, concomitant with issues of inter-and transdisciplinarity, the practice of social science research in the sustainability domain forces consideration of the tension between specialisation and integration. Working across disciplines on large, complex, applied problems requires both rigour and scientific depth, on the one hand, and accommodation and compromise, on the other. This tension often plays out at the expense of the social sciences and social scientists, but it does not have to. Even within research programs that have solely or largely a social science flavour, these tensions exist, and disciplinary hierarchies and jealousies imported from academe have to be resisted. The imperative to mobilise social change for greater sustainability requires these tensions to be resolved, both piecemeal (research project by project) but also institutionally and organisationally. Social scientists must take the lead in doing so. None of this requires the development of a unified or integrated social science. That is likely to be a chimera – the plurality of the social sciences is, and should be celebrated as, a virtue.

Finally, and arguably most importantly, social science research in the sustainability domain must develop a stronger orientation to action. Many of the chapters in this book hint at and skirt around the translation of science into action, but few directly engage in social action and social change. This largely reflects the special social position of a national publicly funded science organisation being seen to engage in action for social change. At the least, social science research can enable, if not directly propel, change towards more sustainable social arrangements.

Institutions react naturally when a problem is well understood, when there is something that can be done about the problem and there is a willingness to act. This book presents encouraging examples of social sustainability science but is also level-headed about the limitations and barriers to sustainable development.

References

Kates RW, Clark WC, Corell R, Hall JMI, Jaeger CC, Lowe I, McCarthy JJ, Schellnhuber HJ, Bolin B, Dickson NM, Faucheux S, Gallopin GC, Grübler A, Huntley B, Jäger J, Jodha NS, Kasperson RE, Mabogunje A, Matson P, Mooney H, Moore B, O'Riordan T, Svedin U (2001) Sustainability science. *Science* **292**(5517), 641–642. doi:10.1126/science.1059386

2

Why do we need a sociology of society's natural relations to inform sustainable development?

Heinz Schandl

The discipline of sociology has been remarkably unprepared for the new challenges of sustainable development, especially around the importance of natural resources in human development and social and cultural evolution. In recent decades, interdisciplinary concepts of social-ecological systems and social (industrial) metabolism have emphasised how social and natural systems interact and perhaps co-evolve in history. They have been, however, mostly ignored by the mainstream of sociological thinking. The focus on the society–nature interface has successfully established new concepts that are non-deterministic and hence have gained recognition in both the social and natural sciences. The social aspect of social-ecological systems and the historical and institutional dimensions have received less attention, creating a vacuum which has been occupied by sometimes oversimplistic approaches, often from outside the social sciences, to understanding social systems. Employing the theoretical contributions of the German sociologists Niklas Luhmann and Jürgen Habermas, this chapter contributes to refocusing on the sociology of society's natural relations. It looks at the contributions of the two great theorists for understanding inertia and agency in social-ecological arrangements and the degree to which modern society has created non-adaptive socio-economic practices that may threaten the natural resources and ecosystems that modern society and human well-being depend upon. A good understanding of the social, institutional and political circumstances that determine society's capacity for sustainable development, which is both inclusive and environmentally sound, will be an important contribution to the literature on globalisation and global change and may create a more realistic picture of the intervention points in an increasingly global society. Achievements in theorising the society–nature interface have not yet been complemented with an equivalent theory of social systems.

Sociology and the environmental problem

The interest of sociology and the social sciences in the environmental and ecological conditions of social life and in connections between the social system and its natural environment started in the early 1980s with Catton and Dunlap's plea for a new ecological paradigm (Catton and Dunlap 1980), that emerged in response to the growing recognition of environmental problems faced by modern society. Society had started to feel the effects of environmental change. Climate change, depletion of natural resources, pollution and waste have become important and increasingly converging features of society's environmental pressures and impacts. Since the 1980s, the prevalent question has been whether sociology can respond to environmental problems in its present state of theoretical development. The foundation of sociology, since its establishment as a scientific discipline in the 1850s, has been the fundamental division of labour between science and social science. The first is for nature and natural laws and the latter for social institutions and people. To be able to address the fundamental environmental problems faced by global society, sociology will need to relax its boundaries, as indeed will the sciences (Dickens 1992). Addressing environmental problems by broadening the theoretical base of the social sciences and creating an interdisciplinary analytical framework has been challenging, and sociology has left a vacuum that has been filled with often simplistic approaches to conceptualising society's natural relations and the connections between social and ecological systems.

In this chapter I argue that the environmental issues that emerged in the 1970s caught sociology unprepared theoretically. Conceptual attempts that aimed to broaden the framework to an interdisciplinary theory of society and nature have been prominent since the 1990s but have largely been ignored by mainstream sociology (Redclift 1999). Interdisciplinary social-ecological research has tried for two decades to change the dominant paradigm, with little success. This has meant that reintegration of sociological theory into the interdisciplinary framework has also been lacking. The aim here is to use the rich theoretical positions of two main theorists – Jürgen Habermas and Niklas Luhmann – as a starting point to delineate the sociological aspects of an integrated theory of society and nature.

One promising variant of social-ecological theory and thinking was developed by the Vienna School of Social Ecology (Newell and Cousins 2015), which has focused its conceptual framework on the interrelationship of social and natural systems (Fischer-Kowalski and Erb 2006), resulting in an advanced theory of social metabolism and colonisation of nature (Fischer-Kowalski and Haberl 1998) and numerous historical and current case studies that have informed environmental policy debate over many decades (Fischer-Kowalski *et al.* 2011). The framework focuses on the hybrid nature of society between natural and social realms (Fischer-Kowalski and Weisz 1999) but remains superficial in furnishing the social domain of the social-ecological system. The authors make reference to Luhmann's theory of the functionally differentiated, self-referential character of modern social systems but do not provide further detail on how Luhmann's comprehensive theory of the social system and its subsystems can be used in the sociological analysis of society's natural relations.

German sociologist and historian Rolf Peter Sieferle has contributed to the Vienna School's theory development, with his own social-ecological model of interaction between nature, population and culture (Sieferle 2011). While nature and culture are viewed as a system in Sieferle's conceptual framework (Sieferle 1997), population is seen as the mediating factor between the two. Only through population is culture able to have a physical influence, either biological or technical, on nature. In the process of labour and by using technologies, artefacts (e.g. houses, roads, bridges and factories) are built; according to Sieferle, these have a physical representation in the natural world and a symbolic representation in culture. Biological and technical interventions into natural systems cause changes in the ways those systems function, which impacts on population in often unintended ways. These impacts are then represented in a specific way in the various cultural subsystems as scientific insights, social conflicts, economic costs, political campaigns, aesthetic innovation or a deficit of religious meaning.

Disturbances that are caused in the relationship between population and nature are not represented directly in culture but are received as irritation and create disorder which can be dealt with in manifold ways. A reaction that relates represented effects to the correct causes is only one, and not even the most probable, reaction.

An alternative approach to understanding social-ecological systems comes from the Institute for Social-Ecological Research in Frankfurt. It has based its conceptual framework on a specific reading of the early Habermas and the Frankfurt School more generally, aiming for a social-ecological theory of society, people and nature (Wehling 2002) that promises insights into the functioning of modern society in broad domains of provision of services, water, mobility and housing and wishes to inform sustainability policy.

I believe it is fair to say, despite the important contributions mentioned above, that the potential contribution of the sociological literature remains underexploited; an all-encompassing theory of social-ecological transition and the potential for social processes to be guided towards a sustainable future require further development.

It has required substantial intellectual effort to establish a social-ecological systems framework and to acknowledge that the relationship of society with nature needs to be seen as a social not an individual relationship. This social-ecological framework also opens avenues for reintegration of sociological theory into the framework aimed at in this chapter.

Socio-cultural life reproduced through processes with outer and inner nature: the contribution of Jürgen Habermas

Why has modern society increasingly put its own environmental conditions at risk? Pressure points of climate change, water and food security, and supply security issues of many strategic materials including metal ores and fossil fuels, are converging rapidly (Weisz and Schandl 2008). To unpack this question we require theory that acknowledges the biophysical underpinnings of social activities and the potential for non-adaptiveness of society to its biophysical conditions. The work of the German sociologist Jürgen Habermas is a good starting point in this regard. In his early work *Legitimation Theory* (Habermas 1973), Habermas analysed multiple crisis tendencies that have formed in late capitalist societies and occur at a systemic level in the economic, political and socio-cultural system. Influenced by Marxist theory, he viewed socio-cultural life as reproduced through processes with outer and inner nature. According to Habermas, any social system would establish a relationship with its natural environment and would appropriate nature, and natural resources, in the production process. He argued that this relationship between the social system and the natural environment (also referred to as outer nature) is maintained through instrumental actions and technical rules. This is essentially Marx's concept of the labour process (Marx 1887), which he described as a process between humans (society) and nature in which society organises and satisfies its physical demands. According to Marx, it is within the labour process and by application of technology that human society organises the essential supply of raw materials and energy (i.e. social metabolism) for production and maintenance of people and artefacts.

For Habermas, building on the insights of the Frankfurt School (Horkheimer and Marcuse 2002) and the work of Norbert Elias (Elias 1969, 1997), the specific relationship between the social system and nature requires a corresponding relationship between the social system and people (and what he calls their inner nature). Establishing this correspondence between outer and inner nature is secured in the process of socialisation through communicative action (see Fig. 2.1) and the establishment of valid norms. The

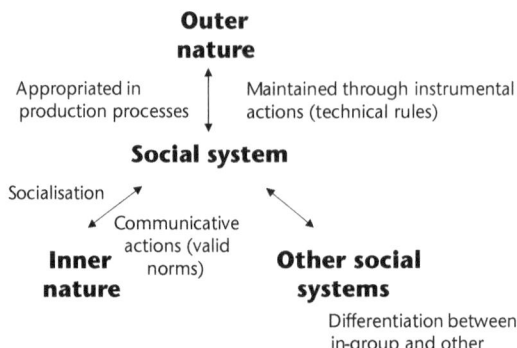

Fig. 2.1. Socio-cultural life reproduced through processes with outer and inner nature.

socialisation process, according to Habermas, guarantees a close match between inner nature and processes in the natural environment. Habermas also referred to social change and history as a process of cultural evolution, in which the relationship between society and nature becomes increasingly complex, institutionalised and removed from human experience and influence.

He described the process of cultural evolution as evolutionary decoupling of economic, political and administrative systems from lifeworld (*Lebenswelt*) driven by functional differentiation and the evolution of modern control media (see Fig. 2.2). In modern society, more and more processes are determined by the rules of the economic and political system which operates, disconnected from the everyday lives of people in communities, neighbourhoods and households. This dominance of the political economy over everyday life has resulted in a process of alienation of people from nature, their relationship to work and from each other. In a romantic sentiment, Habermas lamented the expropriation of lifeworld through the anonymous incentives and power structures of the political economy and its administrative apparatus, which he interpreted as a social domain in which consensus has been replaced by empirically motivated commitment to the incentive of money and the deterrence of power.

This evolutionary move away from the steering arrangements prevalent in the lifeworld where people are connected to certain places, ecosystems and natural resources and in which they organise their relationship with nature, to a hegemony of large socio-technical systems for the provision of essential services such as food, energy, water, housing and mobility, creates a fundamentally new relationship between society and nature. Under the new regime of socio-technical systems (Geels 2004) that has developed hand in hand with modern society, no single individual has command over or influence on the way in which

	System			Lifeworld		
Functional differentiation	Economic system	Political–administrative system	Differentiation of domains	Science	Morals	Art
Control media	Money (incentive)	Power (deterrence), laws	Communication media	Reputation	Value consensus	
Motivation for acting	Empirically motivated commitment substituting for consensus		Motivation for acting	Rationally motivated confidence as a result of previous agreement		

Fig. 2.2. Evolutionary decoupling of system and lifeworld.

essential services, and hence the metabolism between society and nature, are organised. Appealing to individuals to change their behaviours to achieve better environmental outcomes is no more than a sterile reminiscence of a time when people actually operated at the society–nature interface.

In his later work *Theory of Communicative Action* (Habermas 1981), Habermas moved away from the Marxian framework of legitimation theory and turned his attention to the notion of a public domain to explore how a rationalised lifeworld could influence the momentum of a differentiated functional economic and political system (Habermas 2006). In situations that are perceived as crises, such as climate change, pollution or resource depletion, it is possible and even likely that issues that are identified by lifeworld agents who first recognise them can lead to mobilisation of discursive action, initially in specific fora and eventually transported by mass media, making them feature in public discourse and enter the domain of political debate (Habermas 2005).

Therein lies, for Habermas, the potential to heal the fundamental legitimation crisis of modern society through re-empowerment of the sphere of lifeworld achieved by participatory democratic action and a de-legitimation of the abstract powers of money and political power exerted by detached politico-administrative systems in a financialised economy.

Certain aspects of Marxian and Habermasian theory have been employed and adapted to establish modern systems approaches to society–nature interrelations that address the fundamental reliance of any social formation on natural resources and the adaptive capacity of ecosystems. This focus on the society–nature interface, using the interrelated notions of social metabolism and colonisation of nature, has been developed by the Vienna School of Social Ecology.

The Frankfurt Institute for Socio-Ecological Research has developed its own approach to society–nature interrelations based on Habermas' early concepts and with strong references to the Frankfurt School. The approach has a stronger link to the notion of systems of provision and includes an analysis of institutions and infrastructure involved in providing essential services to society, but with a less elaborate conceptual framework of metabolic interactions between social systems and nature than that of the Vienna School of Social Ecology. The Frankfurt Institute of Social Ecology attempts to develop a critical theory of society's socially structured relationship with nature and ultimately a social theory of the environmental crisis that modern society faces. For that purpose, the analytical framework of the Frankfurt School and the early Habermas, before his linguistic turn, appear to be most valuable to constructing theory that includes society, nature and people and is able to interpret social processes in terms of both symbolic and material activities.

Functional differentiation, self-reference and operational closure: sociological systems theory à la Niklas Luhmann

While the theory of Habermas has often been used to empower social movements and agents of deliberative democracy to improve and repair the dysfunctional relationship between system and lifeworld, Niklas Luhmann's sociological systems theory started with a very different conceptual basis and a different set of questions. Luhmann is certainly the most influential German sociological theorist, and his sociological systems theory has provided the most convincing alternative theory of society to Habermas' theory of communicative action. It is fair to say, though, that he has been less favourably received by the Anglo-Saxon research and policy community. This is related to the political philosophy of Anglo-Saxon societies, which is fundamentally different from the European tradition

(Pusey 1987). While the former focuses on the individual, the latter puts greater trust in collective action and social organisation.

From Luhmann's point of view, Habermas overrates the potential of communicative rationality and underestimates the significance of complexity which is central to his systems theory of society (Luhmann 1996). Complexity, in Luhmann's theory, refers to the abundance of potential experiences and actions of which, in any specific situation, only a few can be realised. The range of experiences and actions that can viably be chosen is always extremely small in comparison to the number of alternatives. Complexity hence forces selection.

In Luhmann's theory of society, the ongoing process of cultural evolution refers to the fact that over time society has formed different strategies for social differentiation, allowing increasing internal complexity and enabling more sophisticated relations with complexity in the environment. The major evolutionary achievement of modern society over traditional society is that of functional differentiation into major social subsystems, each of which fulfils a certain function within and for society, on which other subsystems can rely. In a grand division of responsibility in modern society, the economy is responsible for the regulation of scarcity, the political system generates and legitimises collectively binding decisions, the legal system stabilises normative expectations, and science generates new knowledge. In contrast to Habermas, Luhmann makes no value judgements about whether this process of social differentiation is positive or negative. Rather, he uses it as an analytical framework to ask a question which is important to our topic: how, and with what likelihood of success, can modern society respond to humanity's unfolding self-induced ecological crisis?

Another important difference in theory construction is that Luhmann conceptualises these subsystems of society as systems of communication. He argues that each subsystem creates its identity around a specific code which ultimately allows us to identify the system to which any communication belongs. For the economy, the code is about ownership or not owning, and being able to pay or not being able to pay. For the political system, the code is constructed around having or not having power and for science it is about true/false (Table 2.1). There is no third option in the code of any subsystem other than the binary distinction which serves as a filter to identify and relate communication to one of the subsystems.

Every communicative action has two aspects, of containing information and of being coded as belonging to one of the subsystems. The subsystems are conceptualised as self-referentially closed, which refers to the fact that they interpret information according to their specific operations and understanding. They operate according to their code which enables them to maintain their identity even in the case of changes in the structure of the subsystem. Each subsystem establishes programs and creates roles which further stabilise expectations for future communication or behaviour.

The main price modern society pays for its internal complexity is the rising interdependence between subsystems. This is often overlooked, meaning that important phenomena in the environment may remain unobserved. This conceptualisation of social subsystems of communication that interpret information according to their experience and identity makes communication and the exchange of important information or observations about important factors in a society's environment between subsystems more likely to fail. One way to overcome gaps in communication between subsystems is structural coupling, that is, a way in which subsystems are linked and hence can function as cooperating parts of the overall social system. Examples of structural coupling between the legal system and the economy include property rights and contracts; political and legal systems are structurally coupled through constitutions; taxes and levies couple the economy and

Table 2.1. Functional differentiation and self-reference as a characteristic of modern social systems

	Economic system	Political system	Legal system	Science system
Function	Regulation of scarcity	Establishment of collectively binding decisions	Stabilisation of normative behavioural expectations	Generation of new knowledge
Medium	Ownership/money	Power	Law	Truth
Code	To own/not to own; able to pay/not able to pay	In power/not in power (government/opposition)	Justice/injustice	True/false
Program	Investments, purchases	Political programs	Laws, precedents, contracts	Theories and methods
Communicative operations	Transactions	Political decisions	Legal practice, jurisprudence	Publications
Examples for structural coupling	Taxes and charges as structural coupling with the political system	Constitution as coupling with legal system	Ownership and contract as coupling with economy	Scientific consulting as coupling with the political system

the political system; and scientific consultancy is an example of structural coupling between the political system and science.

For Luhmann, in contrast to Parsons (Parsons 1977), structural coupling does not necessarily result in integration of society and does not follow any normative principle. He does not assume that structural coupling results in harmonious development of subsystem relationships. For Luhmann, many interdependencies are in fact left unattended, hence communication between subsystems may occasionally be dysfunctional. Moreover, modern society has no single individual function to observe the interface of society with nature. This gap is filled by observations of different subsystems which interpret problems that may arise at the society–nature interface. For example, natural resource depletion or climate change is a problem of the ability to pay and to financially compensate for risks as would be the case for economic communication, the legal system observes these issues from the perspective of lawfulness or unlawfulness, and the political system interprets the issue from the point of view of maintaining political power and government. Differently from the other subsystems, science may debate whether the assertions of an environmental problem are true or false.

The communication gap that exists in society is also replicated in practice when policy agencies responsible for economic affairs, legal matters and science fail to connect their programs and coordinate their responses to an environmental problem. An additional challenge for modern society is the fact that it has not developed an understanding of itself, that is, of society as whole and a notion of what integrates society. Leaving open the overarching identity of what constitutes modern society as a whole is very powerful for dealing with complexity in so far as it frees up the independent self-regulating structuring and steering capacity of subsystems (Luhmann 1997a).

This lack of an ethical position and overall integration, however, means that many values that exist in social communication (or discourse) cannot be decided upon and the debate and competition over appropriate values often degrades to opinion and moral judgements which are not based on ethical arguments (Luhmann 1989). The lack of overall integration, and the plurality of observations that count, is both a strength and a weakness of modern society and needs to be taken at face value when we analyse the means modern society possesses to confront and solve global sustainability problems.

It needs to be mentioned that the institutions of modern society co-exist with other forms of social integration which are hierarchically structured or are structured around kinship relations. This phenomenon of co-existence of different forms of social integration is even more pronounced in developing countries, where the modern state and its institutions usually co-exist with customary law (McCarthy 2014), creating additional complexity for managing society's natural relations.

Each social subsystem, moreover, has a different understanding of time (Luhmann 1979). The economy has a certain understanding of how long it may take for an investment to yield profit, the political system orients its understanding of time with regard to the electoral cycle, in science an investment into research may take many years before publication of new insights, and legal conflicts may be fought over many years as well. All of these processes, however, operate at altogether different timeframes compared to major environmental problems. There are mismatches between subsystems of society, but also between the social understanding of time and time in society's natural relations and in nature.

Each subsystem resonates to different phenomena in its environment and cannot be steered or guided by the actions of other subsystems. A system, or organisation, can only be irritated by communication that aims to change its capacity to resonate to new phenomena through new information that has so far remained unobserved. For Luhmann, an important feature of modern society, and perhaps of any social arrangement, is its inertia which is based in the self-referentiality and operational closure of each subsystem. This explains why it is hard to change social arrangements even if it is felt that a problem is well understood. One important feature of his systemic approach to society is that the system needs to be understood in its entirety if interventions that plan to change the fundamentals of social organisation are to have a good likelihood of achieving the desired outcomes (Fischer-Kowalski and Rotmans 2009). Chance and failure prevail even if best intentions have guided the intervention.

One important feature of Luhmann's theory, and perhaps the aspect hardest to digest for many social and natural scientists, is the theoretical decision to situate people in the environment of the social system. Social systems rely on communication and establishing a boundary towards an environment. For Luhmann, social systems do not consist of people and actions but of communications. This is a very consequential decision: it allows Luhmann to situate everyday life interactions at the periphery of the social system and hence consider them meaningless and irrelevant for social discourse. It establishes communication as an independent feature of society that follows its own rules. People drive cars, they get drunk and may get married, they have kids and rent a house. But none of these activities really matter for social discourse. They are the noise that oscillates over the broader underlying trend of socially accepted expectations of how people communicate, consume and spend their time, which is regulated and determined by larger discursive patterns and structures at the level of social subsystems.

Social scientists who favour an approach associated with individual choice and human agency as the main driver of social change have criticised the concept of excluding people from society, claiming that Luhmann takes a reactionary stance with regard to people's potential to change their own circumstances and contribute to positive social change (Leydesdorff 2000). Natural scientists who have viewed people as a main factor of disturbance to well-functioning ecosystems have also tended to view them as individual agents, and have often used a simplistic attitude–behaviour model to argue that attitudinal change is required for individuals to change their behaviour to achieve better system outcomes. It is proven that there is only a weak relationship between attitude and actual behaviours (Shove 2010), and little evidence that the amount of change required to put global society on a sustainable development path could result from aggregate individual action.

For Luhmann, in essence, society can be understood as a mechanism that reduces complexity to such a degree that expectations, and expectations of expectations, can be solidified to reduce the pressure of making decisions for individuals and groups of people.

The level of abstraction of expectations rises at different levels of social organisation. At the person level, expectations are usually closely related to previous experiences with that person. If expectations are addressed to roles then it is, in principle, irrelevant who performs the role. There need not be a previous relationship and related experience, to relate certain expectations to someone who fills a role. A third level of abstraction is reached when roles become exchangeable. This is the case for programs such as investment programs of large corporations, research programs of scientific institutions or planning of public investment for transport infrastructure or urban development. In such cases, program objectives and implementation can be identified without specifying in advance which roles in an organisation are responsible for implementation (Luhmann 1997b).

At the highest level of abstraction of expectations, the need to link expectations to a specific context in which expectations may be realised is not necessary. Luhmann speaks about general considerations for specific actions that are preferable to others, i.e. values. Programs can be discussed, evaluated and changed by applying certain values. There may be discussion, for example, around whether it is better to grow economically at the cost of environmental degradation (in order to become rich enough to clean up the damage) or if it is preferable to take a more benign social development path which simultaneously looks after the environment. Conflicts about which paths to choose are difficult to settle because modern society has no overriding principles that enable conflicts between values to be solved. It is important to note that, for Luhmann, values are not a property of individuals but are situated at a higher social level of abstraction of expectations. Because of this, societal values can change without affecting roles or the identity of individuals. Values can change to reflect social change without unravelling expectations at every level of social organisation. In a similar vein, roles can be filled by different people, without affecting the implementation of programs or the larger value set of a society.

The evolutionary advantage of a distinction between people, roles, programs and values as levels of abstraction to which expectations are linked lies in establishing relatively independent variability at each level. Through this, modern society gains greater flexibility compared to traditional society because changes of expectations at one level of abstraction are not necessarily strongly linked to changes of expectations at any other level of abstraction. They are hence easier to achieve. It also enables a larger number and more diverse set of expectations to be institutionalised, which again raises societal potential for building structural complexity and dealing with greater complexity in the environment. However, it also affects the ability of society as a whole to make binding decisions. In the end it remains an open question for modern society: who is in charge of important social decisions? While the political system may appear to have a monopoly on governing society, most evidence points to a multifaceted interplay of different social subsystems providing steering capacity to society (Lemos and Agrawal 2006). The positive side of this is that coalitions for specific social agendas are easier to form since they do not necessarily require integration into an all-encompassing view of the world.

Going beyond actors and agency: a truly sociological approach to society's natural relations

Addressing the very serious ecological crisis that has unfolded globally (Steffen et al. 2015) requires a social theory and analytical frameworks that explain processes of social change and how they are related to changes in the environmental relations established by certain social configurations. Such a theory, based on sociological thinking, enables us to explore the self-endangerment of modern society and identify means by which modern society might address the environmental crisis.

Such a theory needs to depart from an oversimplistic and pre-sociological understanding of individuals who make choices to maximise utility; those choices are guided by motives and attitudes that favour certain behaviours and the idea that if attitudes were to change, this would result in more environmentally benign and responsible behaviours (Shove 2010). This means acknowledging that even a biographical single person represents a social category because they are determined by their life context (Wehling 2002). The way individuals consume, spend their time and make political decisions needs to be seen as a process of social choice rather than individual choice where people pick from pre-determined, socially acceptable ways of leading their life (Duesenberry 1949). We have also seen that environmental relations in modern society are organised in larger socio-technical systems (Geels 2004) such as the energy system, the water supply and sewerage system, the transport system and the food supply system. They consist of specific institutional arrangements, and require large infrastructure that depends on large public and private investments to provide the supply and distribution of natural resources. They can be seen as structural coupling between society and nature in a very technical sense and are embedded in the processes of labour and technology.

These systems need to be represented in social communication and are seen as an issue of cost and rent-ability in economic communication, of rules and regulations in legal communication, and of legitimacy and fairness in political communication. The ways such services are provided to society are determined by broader social values such as equal access, and are rolled out through specific programs and organisations. Most importantly, they are beyond the influence of single individuals and individual choice.

A similar argument can be made for consumer behaviour. We identify this as a social phenomenon with socially formed expectations and aspirations which are filtered by the class structure of society and realised through available supply of housing, mobility and food options and a variety of products that households and individuals can consume. It can be argued that once a set of fundamental decisions has been made, including where to live, whether to marry and have kids, and what kind of work to engage in, a lot of other decisions unfold quite naturally (Duchin 1998). Does the place where someone lives have public transport and shops within walking distance or do people need to rely on private transport? Can distances be met with a bicycle or do people need to go by car? The social position of a person and household will further determine which practices people engage in, which consumer products they choose to furnish a certain way of living, and how they spend their time. Complexity is reduced by expectations, and expectations of expectations, to such a degree that people can manage their lives and do not have to capitulate in the face of never-ending possibilities and the constant pressure of needing to make a choice. In modern society, despite the notion of personal freedom, choice is reserved for domains which matter little for the functioning of society as a whole or for the environmental outcomes of social organisation, such as the amount of greenhouse gases emitted or the required amounts of electricity.

It is hence irresponsible for policy-makers to put moral pressure on individuals for environmentally responsible behaviour when in fact policies and investments that introduce change at the systems level can drive most of the change required to put society on a path to sustainability (Shove 2010).

There remains an open question: which social subsystem is responsible for and will guide the enormous changes in our production and consumption systems that are needed to steer global society on a path towards sustainability? In other words, fundamental changes to the way in which food, housing and mobility are provided to the 9 billion people who will inhabit the planet by 2050 are needed. Who will enable and champion such change?

A response in the framework of Habermas would require a democratisation of public discourse, regaining the sphere of lifeworld and enabling communities and local decision-makers to be involved in the planning of cities, industrial systems and infrastructure to achieve the best outcomes for people, communities and the environment. This would mean a substantial loss of power and decision-making in the sphere of the political and economic system enabled by social agency, conflict and striving for rearrangements in the relationship of system and lifeworld. The main advantage of Habermas' approach is a defined relationship of communities of action and place and hence a geographical intuition linked to, and understanding, social processes and how they are expressed in space (Harvey 1973). This allows for a spatially explicit formulation of sustainability issues (see Chapter 6).

Luhmann's functional analytical framework takes a different stance. It assumes that no subsystem of communication has the ability to guide the change that is required, mainly because of a lack of resonance to the important phenomena that can be summarised as global society's ecological crisis (Luhmann 1989). In addition, important subsystems and independencies remain unattended, which results in a distorted discourse or an absence of discourse in society as a whole. According to Luhmann, the most promising interventions are those which raise the resonance of subsystems to important phenomena, allowing them to adjust their programs while maintaining their identity, making use of the flexibility that is offered by modern, functionally differentiated social arrangements. Any social system observes certain aspects of its environment, i.e. it resonates to certain phenomena and ignores other aspects, which are deemed less important. The notion of resonance implies that social systems cannot be directly influenced or controlled; they choose to react to information according to their own interpretation.

Luhmann insists that there can be too little but also too much resonance, and that it is important to meet the bandwidth at which institutions can respond. This speaks against alarmist as well as overoptimistic contributions to social discourse and asks for a middle ground; a new form of communication of the global environmental crisis beyond optimism and alarmism, laying out the substantial influence that well-designed policies and the social choices that enable them can have on overall sustainability outcomes (Hatfield-Dodds et al. 2015).

Conclusion

It is unclear what the future of sustainable development is going to look like. Will the next big social transition lead into a new stable social-ecological configuration, at the global scale, that we will describe as a sustainable society – in contrast to the current unsustainable industrial society model? The global agreement on sustainable development goals (Griggs et al. 2013) for the world suggests a strong policy-making focus on steering the society–nature relationship towards sustainability. The inherent tension between the focus on economic and human development and the integrity of environmental systems and natural resources is stark. In the conceptual framework of the Frankfurt School, the project of Enlightenment and reason that started in the 18th century would need to become reflexive to avoid a self-directed process of Enlightenment and instrumental rationality driving society towards self-destruction (Horkheimer and Marcuse 2002). But how can social choices enable a society–nature relationship that is not structured by domination and exploitation? How can modern society steer society–nature relations to more equitable and environmentally benign outcomes?

Luhmann's response is that the nature of functionally differentiated and self-referential subsystems and the global character of society reduce the chance for ecological communication to succeed. He also, however, opens up an understanding of modern society that allows for focusing on the phenomena of resonance and steering capacity. He describes how the mal-adaptiveness of modern society to nature has emerged historically, and the shape it has taken, but without presenting a strategy to raise society's resonance for its environmental problems.

Both Habermas and Luhmann have an understanding of cultural evolution as a process of differentiation and specialisation in society, somehow driven by co-evolutionary forces. Social relations and the structuration of society have evolved through time and have led to a level of differentiation typical of modern society enabling a higher level of domination (in Habermas' theory) and a greater ability to deal with complexity (in Luhmann's theory). This process, however, has not concluded and it is important to understand where the process of cultural evolution is headed. Will we see further differentiation or its reversal? At this moment, there are many signs that the latter may be the case, with the economic subsystem owning political decisions and information (i.e. the mass media) and influencing legal arrangements (Harvey 2006).

There is also an important countermovement in society, however, with political representation going to independent candidates, crowd-funding of political donations, disengagement with and distrust of established political parties, the existence of new mass information through social networks and the possibility of a new level of citizenry participation and engagement through social media. It does not appear, however, that this has revolutionised the environmental movement or the way we think about the steering capacity of society–nature relations. While climate change, natural resource depletion and biodiversity loss have reached a new crisis level, everyday life decisions are still divorced from these issues and the policy response often remains partial or symbolic. If the political system regains some steering capacity in concert with other social institutions, to jointly govern society's interaction with nature and to guide and implement the social choices that would allow global society to live within planetary means, then there would be a chance for a sustainable society to emerge. This new social formation may come, however, with a set of sustainability problems that are yet unknown.

Such a fundamental transition would be informed by critical, interdisciplinary and empirically oriented social research that takes the intertwined character of social and natural systems as a starting point and develops a truly sociological understanding of the social character of society's natural relations and sustainability needs.

References

Catton WR, Dunlap RE (1980) A new ecological paradigm for post-exuberant sociology. *American Behavioral Scientist* **24**, 15–47. doi:10.1177/000276428002400103

Dickens P (1992) *Society and Nature: Towards a Green Social Theory*. Temple University Press, Philadelphia.

Duchin F (1998) *Structural Economics: Measuring Change in Technology, Lifestyles, and the Environment*. Island Press, Washington DC.

Duesenberry JS (1949) *Income, Saving, and the Theory of Consumer Behavior*. Harvard University Press, Cambridge, MA.

Elias N (1969) *The Civilizing Process*. Blackwell, Oxford.

Elias N (1997) Towards a theory of social processes: a translation. *British Journal of Sociology* **48**, 355–383. doi:10.2307/591136

Fischer-Kowalski M, Erb K (2006) Epistemological and conceptual bases of social ecology. *Mitteilungen Der Österreichischen Geographischen Gesellschaft* **148**, 33–56.

Fischer-Kowalski M, Haberl H (1998) Sustainable development: socio-economic metabolism and colonization of nature. *International Social Science Journal* **50**(158), 573–587. doi:10.1111/1468-2451.00169

Fischer-Kowalski M, Rotmans J (2009) Conceptualizing, observing, and influencing social-ecological transitions. *Ecology and Society* **14**(2), 3.

Fischer-Kowalski M, Weisz H (1999) Society as hybrid between material and symbolic realms: toward a theoretical framework of society-nature interaction. *Advances in Human Ecology* **8**, 215–251.

Fischer-Kowalski M, Krausmann F, Giljum S, Lutter S, Mayer A, Bringezu S, Moriguchi Y, Schuetz H, Schandl H, Weisz H (2011) Methodology and indicators of economy-wide material flow accounting. *Journal of Industrial Ecology* **15**, 855–876. doi:10.1111/j.1530-9290.2011.00366.x

Geels FW (2004) From sectoral systems of innovation to socio-technical systems: insights about dynamics and change from sociology and institutional theory. *Research Policy* **33**, 897–920. doi:10.1016/j.respol.2004.01.015

Griggs D, Stafford-Smith M, Gaffney O, Rockstrom J, Ohman MC, Shyamsundar P, Steffen W, Glaser G, Kanie N, Noble I (2013) Sustainable development goals for people and planet. *Nature* **495**, 305–307. doi:10.1038/495305a

Habermas J (1973) *Legitimation Crisis*. Beacon Press, Boston.

Habermas J (1981) *The Theory of Communicative Action*. Beacon Press, Boston.

Habermas J (2005) Concluding comments on empirical approaches to deliberative politics. *Acta Politica* **40**, 384–392. doi:10.1057/palgrave.ap.5500119

Habermas J (2006) Political communication in media society: does democracy still enjoy an epistemic dimension? The impact of normative theory on empirical research. *Communication Theory* **16**, 411–426. doi:10.1111/j.1468-2885.2006.00280.x

Harvey D (1973) *Social Justice and the City*. University of Georgia Press, Athens and London.

Harvey D (2006) *Spaces of Global Capitalism: Towards a Theory of Uneven Geographical Development*. Verso, London.

Hatfield-Dodds S, Schandl H, Adams PD, Baynes TM, Brinsmead TS, Bryan BA, Chiew FHS, Graham PW, Grundy M, Harwood T, McCallum R, McCrea R, McKellar LE, Newth D, Nolan M, Prosser I, Wonhas A (2015) Australia is 'free to choose' economic growth and falling environmental pressures. *Nature* **527**, 49–53. doi:10.1038/nature16065

Horkheimer M, Marcuse H (2002) *Dialectic of Enlightenment*. Stanford University Press, Stanford.

Lemos MC, Agrawal A (2006) Environmental governance. *Annual Review of Environment and Resources* **31**, 297–325.

Leydesdorff L (2000) Luhmann, Habermas and the theory of communication. *Systems Research and Behavioral Science* **17**, 273–288. doi:10.1002/(SICI)1099-1743(200005/06)17:3<273::AID-SRES329>3.0.CO;2-R

Luhmann N (1979) Time and action: forgotten theory. *Zeitschrift für Soziologie* **8**, 63–81. doi:10.1515/zfsoz-1979-0105

Luhmann N (1989) *Ecological Communication*. University of Chicago Press, Chicago.

Luhmann N (1996) *Social Systems*. Stanford University Press, Stanford.

Luhmann N (1997a) Limits of steering. *Theory, Culture and Society* **14**, 41–57. doi:10.1177/026327697014001003

Luhmann N (1997b) *Die Geselschaft der Gesellschaft* [*The Society of Society*]. Suhrkamp, Frankfurt.

Marx K (1887) *Das Kapital. A Critique of Political Economy*. Verlag, Hamburg.

McCarthy JF (2014) Using community led development approaches to address vulnerability after disaster: caught in a sad romance. *Global Environmental Change* **27**, 144–155. doi:10.1016/j.gloenvcha.2014.05.004

Newell JP, Cousins JJ (2015) The boundaries of urban metabolism: towards a political-industrial ecology. *Progress in Human Geography* **39**, 702–728. doi:10.1177/0309132514558442

Parsons T (1977) *Social Systems and the Evolution of Action Theory*. Free Press, New York.

Pusey M (1987) *Jürgen Habermas*. Routledge, New York.

Redclift M (1999) Sustainability and sociology: northern preoccupations. In *Sustainability and the Social Sciences: A Cross-Disciplinary Approach to Integrating Environmental Considerations into Theoretical Orientation*. (Eds E Becker and T Jahn) pp. 59–73. Zed Books, London.

Shove E (2010) Beyond the ABC: climate change policy and theories of social change. *Environment & Planning A* **42**, 1273–1285. doi:10.1068/a42282

Sieferle RP (1997) Kulturelle Evolution des Gesellschaft-Natur-Verhaltnisses [Cultural evolution of society-nature-relations]. In *Gesellschaftlicher Stoffwechsel und Kolonisierung von Natur* [*Society's Metabolism and Colonization of Nature*]. (Eds M Fischer-Kowalski and H Haberl) pp. 37–53. Gordon and Breach, Amsterdam.

Sieferle RP (2011) Cultural evolution and social metabolism. *Geografiska Annaler. Series B, Human Geography* **93**, 315–324. doi:10.1111/j.1468-0467.2011.00385.x

Steffen W, Richardson K, Rockstrom J, Cornell SE, Fetzer I, Bennett EM, Biggs R, Carpenter SR, de Vries W, de Wit CA, Folke C, Gerten D, Heinke J, Mace GM, Persson LM, Ramanathan V, Reyers B, Sorlin S (2015) Planetary boundaries: guiding human development on a changing planet. *Science* **347**. doi:10.1126/science.1259855

Wehling P (2002) Dynamic constellations of the individual, society and nature: critical theory and environmental sociology. In *Sociological Theory and the Environment*. (Eds RE Dunlap, FH Buttel, P Dickens and A Gijswijt) pp. 144–166. Rowman and Littlefield, Oxford.

Weisz H, Schandl H (2008) Materials use across world regions. *Journal of Industrial Ecology* **12**, 629–636. doi:10.1111/j.1530-9290.2008.00097.x

3

Integration science for impact: fostering transformations towards sustainability

Ro Hill, Cathy Robinson, Petina Pert, Marcus Barber, Ilisapeci Lyons, Kirsten Maclean, Leah Talbot and Catherine Moran

Sustainability science requires new ways of working that integrate across diverse social, economic, biophysical and cultural knowledge, and link local with global concerns. Recent investigations have highlighted that sustainability science also needs to deliver solutions that are transformational in pace, content and scale, to match the rate of global environmental change. Transformations are characterised by more than just incremental adjustments to policy and practice: changes are required to societal ethics, beliefs, values, institutions, relationships and behaviours. Here we argue that integration science approaches that link scientists and stakeholders into equitable partnerships are highly effective in transformative path-generation. Our analysis of eight case examples demonstrates how these partnerships mobilise capacities across cognitive, social, material and normative domains. Key tools for mobilising these capacities include place-based connections, common problem/issue framing, dialogues, visits to country, participatory modelling, negotiating co-produced knowledge outputs, access to sufficient resources (particularly time) and actors whose enduring relationships facilitate cross-scale linkages. Engaging Indigenous knowledge requires approaches that reflect their unique roles and responsibilities. We illustrate how and why mobilising these capacities produces the types of changes necessary for transformative path-generation: changes in ethics (e.g. making marginalised peoples visible); changes in beliefs (e.g. empowering people to believe they can bring about change); changes in values (e.g. shifting towards recognition and valuing of Indigenous people and their roles); changes in institutions and relationships (e.g. new collaborative institutions underpinned by stronger relationships between diverse groups of stakeholders and scientists) and changes in behaviours (e.g. use of language). The case examples also show evidence for integration gaps that lead to ongoing marginalisation of Indigenous knowledge, and our discussion considers the potential for Indigenous research methods to address these gaps. We conclude by reflecting on the importance of social science methods and approaches that increase the ability of diverse groups in society to integrate knowledge and address pressing land and water sustainability challenges.

Global-scale analyses indicate that we have exceeded planetary boundaries on many biophysical parameters (e.g. biodiversity loss and nitrogen cycles), thus degrading ecosystem services. Concurrently, we are failing to achieve just outcomes for many social parameters (e.g. food provisioning, access to water and social equity) (Rockström et al. 2009; Leach et al. 2013; Costanza et al. 2014). Habitat loss, species extinctions, declining water quality and accessibility, and loss of Indigenous languages and traditions are some of the many current sustainability problems that display human–environment impacts and substantial differences in effects due to histories and cultures (Kinzig 2001; Brown et al. 2010; Jacobs et al. 2010). Evidence is accumulating that dramatic, transformational-scale socio-cultural, political and technical changes are required to address such problems, in contrast to the incremental technology or policy changes that addressed previous environmental problems such as point-source water pollution and acid rain (Geels 2010; O'Brien 2012). In this chapter, we argue that integration science approaches that link scientists and stakeholders into equitable partnerships are highly effective in transformative path-generation.

We use the term *integration science* to encompass both 'interdisciplinary' research, which involves transcending the boundaries between research disciplines to create integrated outputs and outcomes, and 'transdisciplinary' research, which involves transcending the greater boundaries that can exist between scientific, practitioner, local, and Indigenous knowledge systems (Tress et al. 2005). In Australia, integration science has been promoted in the environmental domain over the past 20 years. Key programs that have brought researchers, stakeholders and research funders together include Cooperative Research Centres, Rural Research and Development Corporations, National Environmental Research Programs and Centres of Excellence such as the Australian Centre for Ecological Analysis and Synthesis (Campbell et al. 2015; Specht et al. 2015). Similar initiatives have been supported in the environmental domain globally, including the International Human Dimensions Program on Global Environmental Change, Future Earth and the recent Intergovernmental Platform on Biodiversity and Ecosystem Services (Brito and Stafford Smith 2012).

Evaluative analysis of the outcomes of these initiatives has identified numerous benefits such as data-sharing, enhanced productivity, enhanced theoretical clarity and improved conceptual understanding. Practices identified as necessary to deliver these benefits include conduct of working groups, face-to-face meetings, informatics and support for remote engagement, and culturally sensitive interactions among leading influencers from policy, management and science (Lynch et al. 2015). Integration science has also made positive contributions to specific environmental problems globally, including grass pollen allergens (Davies et al. 2015), air quality management (Stahl and Cimorelli 2013), climate change and agricultural production (Podestá et al. 2013), mobilising institutions with scale-dependent comparative advantage (Hill et al. 2015b) and many others (Brown et al. 2010). However, elucidation of the linkages between integration science and sustainability trajectories more generally is at an early stage of development.

Scholarly attention to understanding transformational changes has identified diverse interpretations, and key questions regarding transformation of what and for whom (O'Brien 2012). Both coercive and generative power feature in many transformations. Issue-based social discord and actor agency, particularly of marginalised groups, often emerge and stimulate change. Long time-scale trajectories are required for eventual uptake into societal norms (Hendriks 2009; Hill et al. 2013a; Olsson et al. 2014). Engagements through integration science approaches appear to provide a fertile ground for incubating transformations to sustainability (Pereira et al. 2015). Nevertheless, knowledge gaps remain in relation to how, why and under what conditions integration science approaches support such incubation for transformative path-generation (Fazey et al. 2014; Robinson et al. 2014b).

In this chapter, we draw on eight case studies of integration science, led by social scientists and geographers, which addressed Australian land and water sustainability problems, in order to consider their linkages with transformative path-generation. We analyse examples across climate change, biodiversity protection and natural resource management from local, regional (sub-national) and national scales within Australia, to global scales. We identify how mobilisation of capacities in social, cognitive, normative and material domains results in changes to ethics, beliefs, values, institutions and behaviours. We conclude with a discussion of the relationship of these changes to the path-generation required for sustainability, and reflect on the role of integration science approaches in delivering positive research impacts with societal, environmental, cultural and economic benefits for Australians and the global community.

Key concepts
Integration science and required capacities

Integration science (as we have defined it for this chapter) encompasses both transdisciplinary approaches that link scientists and practitioners, and approaches that link disciplines, variously termed interdisciplinary, multidisciplinary and cross-disciplinary in the research literature (Rosenfield 1992; Tress *et al.* 2005; Stokols *et al.* 2008). Plurality and convergence in approaches between disciplines is increasingly a characteristic and emerging strength, particularly as the result of transdisciplinary approaches to facilitating integration across disciplines (NRC 2014). The concept of co-productive capacities has recently been identified as a useful new lens through which to examine the skills necessary for tackling these problems (Wyborn 2015). Co-productive capacity is understood to mean the capability to operationalise relationships between scientific and public, private and civil institutions and actors to bring about scientifically informed social change (van Kerkhoff and Lebel 2015). Effective integration science clearly requires such capacities, which Wyborn (2015) classifies across four domains: social (communication, mediation, translation), normative (place attachment, leadership, vision), cognitive (credibility, salience, legitimacy) and material (funding and cross-scale linkages). We use this framework of four domains of capacities to analyse the insights provided by our case examples (Fig. 3.1).

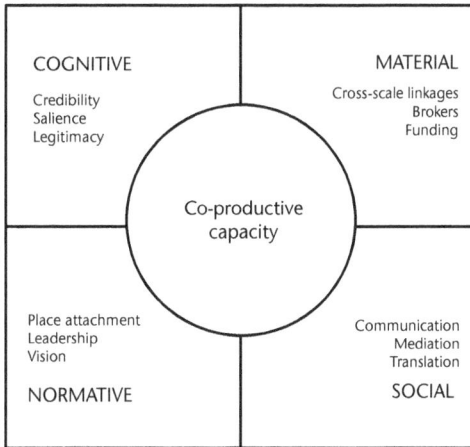

Fig. 3.1. Co-productive capacities framework. Source: Wyborn (2015).

Transformational change and transformative path-generation

Transformational change has diverse interpretations, depending on what is being transformed, why and in whose interests (O'Brien 2012). Here we focus on the concept of 'transformational change' as a deliberate attempt to move towards sustainability globally, living within planetary boundaries while providing socially just and equitable access to food, water, education and other requirements that support human development, capabilities and well-being (Sen 2005, 2013). Such transformation involves a profound and fundamental change in human–environment interactions, which in turn relies on changes of ethics, values, beliefs, relationships, institutions, behaviours and actions (Westley *et al.* 2013). Transformational change shifts away from accommodating change (adaptation) to contesting and creating transformative actions for alternative futures through dynamic, complex and systemic processes that involve different phases over time (O'Brien 2012; Hill *et al.* 2013a). While transformations may not be visible in the short time-frames of contemporary research and practice, or political cycles, transformative pathways for a particular issue (e.g. sustainability, gender equity) can be identified (Leach *et al.* 2010, 2012). Past societal choices can generate path-dependency, constraining the ability to move an issue in a different direction (Olsson *et al.* 2008; Gelcich *et al.* 2010). Nevertheless, path-generation through human agency and creativity, for example through grassroots social movements, can shift these constraints (Djelic and Quack 2007). Understanding transformative path-generation requires knowledge of histories for explaining social structure (how social hierarchies and dynamics have developed), and of geographies for explaining ways in which 'natural landscapes' have been invested with cultural meaning. These explanations and understandings provide a foundation from which to elucidate the impact of interventions such as integration science projects (Câmpeanu and Fazey 2014).

Geography and context of case examples

The case examples presented here range in scale from global to Australian national, regional (sub-national) and local levels. They address sustainability challenges associated with biodiversity, biocultural diversity, climate change adaptation and associated emerging carbon economies, and natural resource management (Fig. 3.2). Table 3.1 details who was involved in the cases, where and when they occurred, and the drivers for the application of integration science approaches.

At the global scale, two case examples address aspects of biodiversity conservation (Table 3.1). Integration science through science-stakeholder engagement was critical to bringing Indigenous and local knowledge (ILK) into the first global assessment undertaken by the Intergovernmental Platform on Biodiversity and Ecosystem Services (IPBES), *Pollination and Pollinators in Food Production* (Potts *et al.* 2016). IPBES aims to include ILK in all its assessments, recognising the value of multiple knowledge systems (Díaz *et al.* 2015). Activities to support the engagement of ILK included: hosting a global dialogue workshop in Panama in December 2014 (Lyver *et al.* 2015); subsequent community workshops in Asia, South America, New Zealand and Africa; gathering ILK resources and screening for materials that met ethical standards; and analysis of materials by an interdisciplinary team of scientists for inclusion in the final assessment.

The social-ecological systems analysis of impediments to achievability of the global biodiversity targets, known as the Aichi Targets (Secretariat of the Convention on Biological Diversity 2011) was achieved through conceptual modelling, a global workshop and

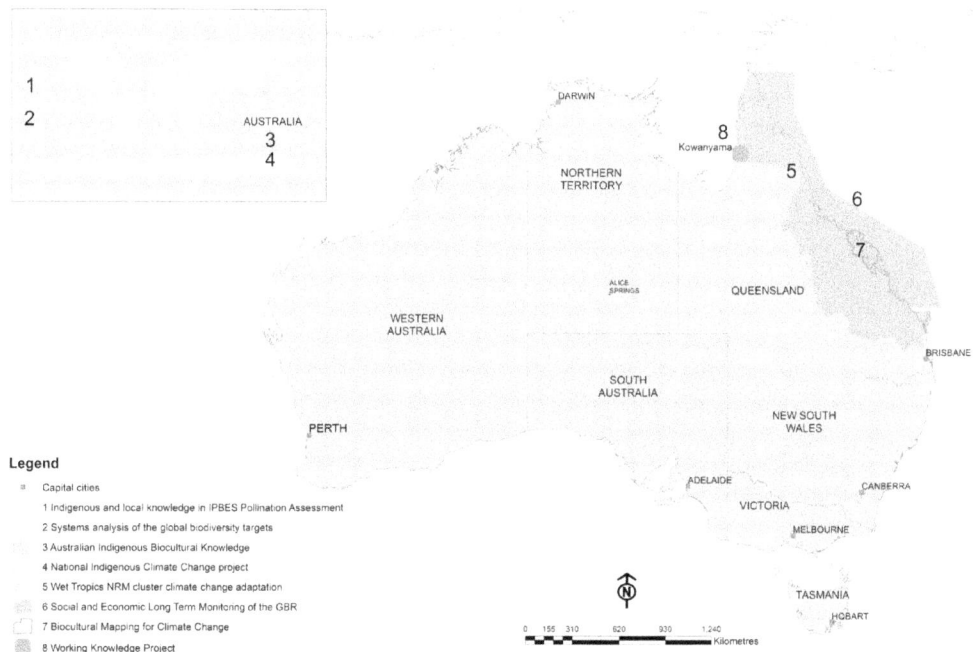

Fig. 3.2. Location of case examples: two each at global, national, regional and local scales.

ongoing interactions to review workshop products, by a team of scientists with diverse disciplinary backgrounds. This analysis was driven by an interest in understanding if and how decoupling between political, economic, environmental and social policy domains explains why biodiversity continues to decline despite decades of activity under the Convention on Biological Diversity (Hill et al. 2015c).

At the national scale, one case example addresses aspects of Australian Indigenous biocultural knowledge (IBK). The IBK case involved scientists and Indigenous people working together through two workshops, and ongoing interaction by email, telephone and documentary review. A spatially located compilation was made of documented Australian IBK, to address knowledge gaps about the availability, extent and location of IBK resources, which are increasingly sought for national and global assessments (Ens et al. 2015; Pert et al. 2015).

The other national case example, National Indigenous Climate Change Project (NICC), occurred through a collaborative forum, established by Indigenous leaders, that facilitated workshops and policy dialogue among corporate Australia, Indigenous peoples, scientists and other experts (Robinson et al. 2014a). NICC responded to the opportunity to assist Indigenous consultation and participation in Australian climate mitigation legislation and policy, and ensure their rights and participation were advanced. This included negotiating Indigenous benefits from payment for ecosystem service (PES) schemes established as part of carbon offset agreements (Robinson et al. 2016a, b)

At the regional scale, there are two case examples. The Social and Economic Long-term Monitoring Program (SELTMP), conducted across the Great Barrier Reef (GBR), involved science-stakeholder engagement through sectoral working groups (e.g. fishing, tourism, agriculture, recreation, ports and shipping). An interdisciplinary team of scientists

Table 3.1. Case example teams, activities, locations, timing and drivers for integration science

Case title and key sources	Team	What happened	Scale, place and timing	Drivers for integration science
1. Indigenous and Local Knowledge about Pollination and Pollinators Associated with Food Production (Díaz et al. 2015; Potts et al. 2016)	Transdisciplinary team of Indigenous and local knowledge holders and scientists from parts of Africa, Asia, North, Central and South America, Europe, Australia and New Zealand.	Assessment through literature review; engagement with living knowledge systems through a global call and subsequent workshop followed by community-based activities; information collection focused on practices relevant to pollination services globally.	Global scale of knowledge integration with co-production activities focused in Indonesia, New Zealand, Kenya, Brazil and Panama. Occurred during 2014–15.	Requirement for the integration of ILK into assessments of the Intergovernmental Platform on Biodiversity and Ecosystem Services.
2. Social-ecological systems analysis of the achievability of the global biodiversity targets (Aichi Targets) (Hill et al. 2015c)	Interdisciplinary science team of nine across anthropology, biodiversity, biology, ecology, economics, ecosystem services, governance, human geography, land use modelling, landscape planning, political ecology and social–ecological systems science.	Conceptual modelling of social-environmental-political-economic system interactions to identify impediments to protecting biodiversity and ways to overcome these impediments.	Global scale for analysis. Undertaken 2011–14 through face-to-face meetings in 2011 and 2014 and digital knowledge integration activities in between.	Interest to understand if human–environment interactions explain why biodiversity continues to decline despite decades of activity under the Convention of Biological Diversity.
3. Australian Indigenous Biocultural Knowledge (Pert et al. 2015)	Transdisciplinary team of Indigenous people, ethnobotanists, scientists and Indigenous rangers with expertise in IBK resources from all states of Australia.	Two workshops involving scientists and Indigenous peoples, synthesis of data, establishment of Endnote database, building of interactive website (aibk.info) and production of maps.	National scale using publications from 1843 to present. Study was undertaken in 2012–14.	Interest to address knowledge gaps about the availability, extent and location of resources for IBK; such knowledge is increasingly sought for national/global assessments.
4. National Indigenous Climate Change project (Robinson et al. 2014a)	Transdisciplinary team that brought in policy, Indigenous, corporate and practitioner knowledge to enable Indigenous benefits from carbon offset schemes.	Collaborative forum, established by Indigenous leaders, to provide policy dialogue between corporate Australia, Indigenous peoples and other experts about issues, risks and opportunities associated with climate change and participation in the carbon market.	National scale project conducted in 2010–14.	Opportunity to assist policy formulation relevant to Indigenous people to provide (and be paid for) environmental services that are aligned with Indigenous customary and contemporary obligations to country.

Case title and key sources	Team	What happened	Scale, place and timing	Drivers for integration science
5. Wet Tropics Natural Resource Management (NRM) cluster climate change adaptation (Bohnet et al. 2013; Hill et al. 2015b)	Transdisciplinary team with regional NRM planners and managers, Indigenous peoples, scientists, with biophysical, economic, social science, species modelling, spatial analysis, ecosystem services and practitioner expertise.	Knowledge brokering hub, with dedicated knowledge broker, between two research and four NRM organisations; co-production of reports and fact sheets, provision of data, maps and models from the scientists to the NRM organisations, co-design and delivery of stakeholder engagement workshops and events.	Regional scale study of the Wet Tropics Cluster NRM region, undertaken from 2013 to the present.	Requirement for the integration of climate change knowledge into regional NRM plans.
6. Social and Economic Long-term Monitoring Program (SELTMP) for the GBR (Adger et al. 2013; Goldberg et al. 2016)	Interdisciplinary team of seven: anthropology, biodiversity, fisheries, recreation and tourism management, biology, ecology, economics, ecosystem services, governance, spatial analyses and modelling, landscape planning, political ecology, and social-ecological systems science.	11 sectoral working groups of community, government, research and industry. Survey of over 8000 coastal and national residents, commercial fishers, tourism operators and tourists, including online, face-to-face, telephone and in-depth face-to-face interviews. Analysis and mapping of patterns.	Regional scale study of the GBR and coastal catchments, undertaken 2011–14.	Requirement for GBR managers, industries and communities to understand and monitor the human dimension of change and effectiveness of responses in the GBR region.
7. Biocultural Mapping for Climate Change	Transdisciplinary team of Indigenous local and traditional knowledge holders, NRM practitioner, social scientist and ethnobotanist.	Support for local knowledge recording and biocultural mapping incorporating Indigenous aspirations for climate adaptation and contributions to the regional climate action plan.	Local scale study with Mackay-Whitsunday traditional owners, conducted 2013–15.	Strong interest in and need to strengthen climate adaptation planning through knowledge sharing, inclusive decision-making and use of Indigenous ecological knowledge and science.
8. Working Knowledge Project (Barber et al. 2014)	Transdisciplinary team of social and natural scientists collaborated with Indigenous and local knowledge holders, including community-based natural resource managers, pastoralists and Indigenous elders.	Recovery and synthesis of ecological knowledge through documentary analysis and direct contributions from multiple sources – scientific, Indigenous, pastoral, local NRM and others.	Local scale analysis across three years, focused on a remote but ecologically valuable area of Cape York Peninsula, north Queensland, that is owned and managed by the Kowanyama community.	Requirements of local managers to fill knowledge gaps, and the need to better conceptualise the contemporary knowledge base used by Indigenous land managers.

gathered data through >8000 face-to-face, online and telephone surveys and interviews. SELTMP responded to the need for GBR managers, industries and communities to understand and monitor the human dimension of change and the effectiveness of responses (Marshall *et al.* 2015).

The Wet Tropics Natural Resource Management (NRM) cluster climate change adaptation project involved science-practice engagement through ongoing regular meetings of a 'brokering hub' between researchers and NRM organisations, employment of a knowledge broker, and co-production of materials between practitioners and a science team with diverse disciplinary backgrounds. The project responded to the requirement for the integration of climate change knowledge into plans of the NRM organisations encompassing land and seas of the Torres Strait, Cape York Peninsula, Wet Tropics and Mackay-Whitsunday areas.

Two local case examples particularly focused on the fine-scale knowledge of Indigenous Traditional Owners. The Biocultural Mapping for Climate Change project involved scientists, NRM practitioners and Traditional Owners spending time on country together and sharing knowledge of their historical and current relationship and aspirations in grounded exchanges. The project was driven by the need for the inclusion of Indigenous peoples and their knowledge into climate change adaptation. The Working Knowledge project, conducted in collaboration with Kowanyama Aboriginal Land and Natural Resource Management Office, involved fieldwork at Oriners Station by scientists, Traditional Owners, pastoralists and NRM practitioners. Field results were combined with documentary and scientific data analysis, leading to synthesis of knowledge. The project responded to the need for local knowledge recovery and for the establishment of a contemporary hybrid knowledge platform to support natural and cultural resource management of this ecologically valuable area of north Queensland.

Capacities mobilised in the case examples

Our case examples highlight how integration science approaches in the land and water sustainability domain engage a range of knowledge types (Table 3.2). However, the knowledge types coming together were diverse, triggering different challenges. The number of different knowledge types in each case example is one measure of diversity. The diversity of underpinning philosophies is possibly a more important measure, which we have characterised as levels of contrast versus commonality. Philosophical contrast or commonality can arise from diverse epistemologies including positivism, interpretivism, constructivism, critical enquiry and narrative/holistic Indigenous approaches (Crotty 1998; Wilson 2008; Khagram *et al.* 2010). Philosophical differences can also arise because of diverse ontologies, such as the naturalistic ontology of western science and the totemic ontology of Australian Indigenous knowledge systems (Descola 2014). Issues of world-view construction, interpretive flexibility and competing knowledge claims can be challenging to negotiate in real-world contexts (Van Opstal and Huge 2013). Negotiating the normative boundaries between science and Indigenous knowledge is particularly challenging: in addition to philosophical barriers, efforts to bridge Indigenous and scientific knowledge systems encounter power differences, the absence of a common language and diverse encounters between Indigenous and agency governance and co-governance (Hill *et al.* 2012; Robinson and Wallington 2012). These differences affected how the co-productive capacities were mobilised. All four co-productive capacities identified by Wyborn (2015) were mobilised in the examples (Table 3.2).

Normative domain capacities

The normative domain concerns the underlying values that inspire actors to work towards a common goal. 'Place-based connections' and 'common problem/issue framing', both of which played a role in bringing underlying normative values to the fore, underpinned capacities in this domain. Engagement with Indigenous knowledge also highlighted the benefits of the ongoing presence of Indigenous people on their traditional lands for knowledge integration about land and water sustainability, as well as the challenges of the fit between the holistic and reductionist emphasis in Indigenous knowledge and science respectively.

Place-based connections anchor high levels of commonality and were an evident influence in the GBR, Wet Tropics NRM, Biocultural Mapping and Working Knowledge case examples. A hybrid form of knowledge, termed 'working knowledge', was identified as key to providing an underlying platform of commonality in the Oriners Station case. The choice of this term reflects the contexts in which the knowledge was obtained (through pastoral, Indigenous, NRM and scientific labour), the diverse backgrounds of the project participants, the provisional and utilitarian quality of the knowledge, and the focus on aiding adaptive management of land and water. The working knowledge formulation enabled the synthesis of contributions from different sources – Indigenous people, pastoralists, scientists and NRM practitioners – to be accurately characterised (Barber *et al.* 2014). In the Biocultural Mapping case example, both place-based connections and common normative values around environmental protection anchored commonality:

> *Reef Catchments is about the environment, we're about looking after the environment. Helping create them partnerships. A lot of our cultural sites are on national parks. We're a bit restricted there too because they'll (National Park) say yes, (Traditional Owners) you look after the sites, we (National Park) manage the parks. They (sites) are our historical connection to country. We should be able to be involved in managing and looking after them. That's what we want to do (Traditional Owner).*

SELTMP brought together individuals and communities with strong place-connections with the GBR, through culture, occupation or familiarity. This allowed multiple stakeholders and end-users to come together to express their needs for more information and help to identify critical knowledge gaps in existing social and economic datasets, aimed at formulating a long-term social and economic monitoring program to underpin NRM.

Common problem/issue framing occurred in all the case examples, most often resulting from workshops that brought together scientists and non-academic partners from industry and community, to collate diverse perspectives on the problem domain and to test solutions. Bringing ILK into the global pollination assessment was particularly challenging because of a lack of place-based or common problem-solution frames as anchors for commonality. The focus on pollination as a separate theme, and cause-and-effect relationships viewed by the science teams as central to understanding pollination (e.g. pollen fertilising ovum), were not seen as a useful framing by the community participants:

> *We do not see pollination as a separate theme. Rather that everything – trees, rivers, the wind, even human beings – participates in the process. We cannot separate them (Elmer Enrico Gonzalez López, oral presentation p. 42 in López et al. 2015).*

The quote was from a Guna leader speaking at the Global Dialogue Workshop for the Pollination Assessment.

Table 3.2. Diversity of knowledge types and co-productive capacities mobilised in the case examples

Case example	Range of knowledge types	Level of contrast v. commonality	Normative domain capacities	Social domain capacities	Cognitive domain capacities	Material domain capacities
1. Indigenous and Local Knowledge about Pollination and Pollinators Associated with Food Production	Biophysical, biological, economic, social sciences, Indigenous and local peoples' knowledge about pollination and pollinators.	High level of contrast between dominant positivist science and more holistic Indigenous and local peoples' framing of 'pollination'.	Challenging due to absence of place-based connections or common vision; common issue framing assisted.	Dialogue workshops at global and community scales.	Case study co-production between scientists, Indigenous and local peoples.	Sufficient resources for workshops. Indigenous and Local Knowledge Taskforce provided ongoing support.
2. Social-ecological systems analysis of the achievability of the global biodiversity targets	Ecological, biological, economic and social scientific knowledge about how system interactions affect biodiversity loss.	High level of commonality between scientists with flexibility in positivist and interpretivist approaches and experience in systems thinking.	Common interest in halting biodiversity loss, common problem-framing.	Dialogue workshops at beginning and towards finalisation of project, extended scientific peer review.	Conceptual modelling, extended scientific peer review.	Sufficient resources for international workshops. Stable organisations to support interaction over several years.
3. Australian IBK	Indigenous peoples' knowledge of their living biocultural knowledge systems; spatial, ecological and social science knowledge.	High level of commonality between scientists with flexible approaches, experience with Indigenous peoples, and Indigenous people who are also trained in science.	Common interest in a theme (Australian IBK).	Working group between scientists and Indigenous peoples meeting several times throughout the project.	Co-production of web pages, publications, extended review by Indigenous peoples and scientists.	Sufficient resources for workshops and for developing web pages and associated co-produced papers.
4. National Indigenous Climate Change project	Economic, legal and social knowledge of Indigenous peoples, scientists and business leaders about the benefits Indigenous communities wished to achieve from participating in carbon enterprises.	Mixed high level of commonality between researchers and practitioners trying to find solutions to a common goal. Contrast with finding agreed institutional pathways to achieve Indigenous carbon co-benefit goals.	Common interest in a theme (National Indigenous Climate Change).	Working group and Indigenous Steering Group conducting dialogue workshops, roundtables and forums.	Co-produced roadmap, co-written submissions to policy processes.	Sufficient resources for workshops and forums that brought a high level of skilled people to the task.

Case example	Range of knowledge types	Level of contrast v. commonality	Normative domain capacities	Social domain capacities	Cognitive domain capacities	Material domain capacities
5. Wet Tropics Natural Resource Management (NRM) cluster climate change adaptation	Ecological, economic, cultural and social science, together with NRM practitioners and Indigenous peoples about climate change in north-eastern Australia.	Medium level of commonality, place-based connections between NRM practitioners, Indigenous peoples and scientists, but some contrasting approaches, some highly positivist science and practitioners' practice orientation.	Place-based connections, common interests in climate change adaptation.	'Brokering hub' established as the key interface between the NRM practitioners and the science team, also multiple workshops and forums for engagement.	Knowledge co-production through interactions to produce information sheets and workshop presentations, knowledge broker to engage between scientists and practitioners, extended science and partner peer review.	Sufficient resources for ongoing meetings of a 'brokering hub' over several years, employment of a knowledge broker. Concurrent funding of science consortium and NRM groups to collaborate in the project.
6. Social and Economic Long-term Monitoring Program (SELTMP) for the GBR	Social and economic data and knowledge relevant to diverse groups in the GBR including Traditional Owners, tourists, tourism operators, commercial fishers, catchment industries, ports, shipping, and national and local residents.	Medium level of commonality provided by strong mutual connections to the GBR, contrast high in some cases e.g. between some commercial fishing and community groups.	Place-based connections; common problem framing with industry/community/government to determine monitoring focus.	Sectoral user-groups (e.g. fishing, tourism) workshops and working groups.	Participatory conceptual modelling.	Sufficient resources for large-scale data collection and analysis and ongoing meetings of several sectoral working groups.
7. Biocultural Mapping for Climate Change	Climate science projections, including risk maps of flood, erosion, sea-level rise, adaptive capacity and including NRM and local knowledge, social science, botanical knowledge and Indigenous local and traditional knowledge in the Mackay-Whitsunday coastal region.	Medium level of commonality arising as a result of scientists with flexible approaches, experienced in working with Indigenous peoples, and Indigenous people who are also trained in science. Also common place-based connections between NRM practitioners and Indigenous peoples.	Common interest in a theme (climate change impacts on natural resource management), common framing of the historical trajectory underpinning current conditions.	Camping on country together, mutual exchanges about understandings and experiences of the place and changes experienced.	Co-production of workshop presentations and maps, mapping of potential inundation through sea-level rise and impact on important cultural sites, co-development of sense-making narratives.	Sufficient resources for camps on country, and for bringing multidisciplinary practitioners and Indigenous teams together.
8. Working Knowledge Project	Indigenous, scientific and local knowledge about the ecology, hydrology and geomorphology of a savanna cattle station.	High level of commonality from place-based connection leading to hybrid 'working knowledge'.	Place-based connections, common problem framing.	Formal knowledge sharing and elicitation on country.	Extended peer reviews, conceptual modelling synthesising influences perceived from diverse perspectives.	Sufficient resources for time together on country with science, Indigenous and practitioners' teams.

The Pollination Assessment highlights the need to appreciate this difference, in a key message:

Scientific knowledge provides extensive and multi-dimensional understanding of pollinators and pollination resulting in detailed information on their diversity, functions and steps needed to protect pollinators and the values they produce. Pollination processes in indigenous[1] and local knowledge systems are often understood, celebrated and managed holistically in terms of maintaining values through fostering fertility, fecundity, spirituality and a diversity of farms, gardens and other habitats (Potts et al. 2016, p. 6).

Social domain capacities

The social domain capacities are concerned with the ability to engage with a broad spectrum of stakeholders (Wyborn 2015). Dialogues and visits to country were the key tools used to mobilise social domain capacities. Dialogues supported the operation of multi-agency groups as the key governance driver for the projects. Visits to country and practices of free, prior and informed consent were particularly important for Indigenous peoples.

Dialogues were fostered through working group meetings, face-to-face workshops and digital technologies. The NICC team worked with a policy-level forum, the Carbon Farming Initiative Indigenous Leaders' Roundtable, chaired by the Hon. Mark Dreyfus MP, with key corporate representatives, Indigenous leaders and the NICC project team, to identify key aspects of Indigenous enterprises that could contribute to climate mitigation responses (Robinson *et al.* 2014b). A key premise was that carbon co-benefits may be a negotiated 'end product' of free and fair engagement with Indigenous Australians and should not be acquired by 'stealth' (Robinson *et al.* 2016b). Free, prior and informed consent are critical to ensuring that partnerships are based on mutually respectful, transparent and trusted foundations.

Visits to country with Indigenous peoples and others also occurred, reflecting Indigenous peoples' view of social relationships as extending into relationships with plants, animals and country (Walsh *et al.* 2013). In the Biocultural Mapping case example, visits to country allowed discussions about observations of change, aspirations to look after country and to ground truth mapped western science climate risk assessments. The visits also established the social connections and relationships necessary for knowledge integration:

Well I guess the first thing is actually, having that relationship where acknowledging what is important to us and then trying to work together to have some kind of outcomes, may be projects ... It's just usually when you coming out, if the sites were there, you just feel that connection more, that you still have some sites that are still here. Just makes you feel alive. Yes your culture is still alive for future generation (Traditional Owner).

The Traditional Owners involved in this case example emphasised that the Indigenous way of talking on country is very different from the formal workshop setting:

The way of 'doing business' through the workshop process is very different to the way Aboriginal people work. The emphases of workshop-style collaboration are theory and

---

[1] Indigenous is upper case throughout this chapter, reflecting the Australian preference. However, lower-case is retained in this quote, reflecting the IPBES usage.

talking; for Aboriginal people the standard process of collaboration is of learning by doing things together and sharing knowledge, usually on the ground, on traditional country (Samarla Deshong in Hill et al. 2015a, pp. 46–47).

Cognitive domain capacities

Cognitive capacities concern the processes of generating knowledge that has salience, credibility and legitimacy and then turning that knowledge into action. Credibility derives primarily from technical capability and arguments, salience from the relevance of knowledge to decision-makers' needs, and legitimacy from whether information is generated in fair and unbiased processes that respect actors' divergent beliefs and values (Cash *et al.* 2003). Participatory modelling and negotiating co-produced knowledge outputs, including fact sheets and other documents, were key to cognitive domain capacities. Modelling with Indigenous peoples again highlighted relationships between people and country.

Participatory modelling co-produced knowledge that was technically credible, relevant to the task at hand and allowed all participants to contribute. Several diagrams (conceptually) modelling diverse one-way influence relationships between biophysical and social phenomena were developed in the Oriners Station example (Fig. 3.3). These models brought together a range of knowledge types and highlighted key socialand biophysical linkages. They demonstrate the importance to sustainability of Indigenous residence on the country. A diagram (again conceptually) modelling two-way interactions between social, political, economic and environmental phenomena along six axes was developed for the social-ecological systems analysis of achievability of the Aichi Targets (Hill *et al.* 2015c). A diagrammatic conceptual model for the SELTMP, developed between community, industry and government, was an early outcome from that project (Fig. 3.4).

Negotiating co-produced knowledge outputs occurred in four case examples (Table 3.2). Five co-produced case examples were particularly important for the Pollinations assessment, which included one from each global region associated with IPBES (Africa, Asia-Pacific, Americas and Europe) and another from the Guna hosts of the global dialogues workshop (Hill *et al.* 2016). Co-production occurred through community dialogues

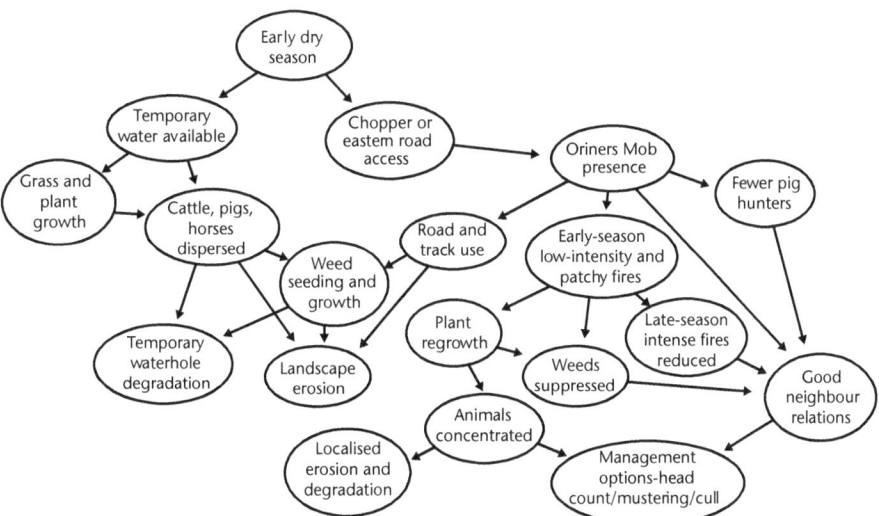

Fig. 3.3. Oriners Station landscape interactions in the early dry season, incorporating residence by Indigenous owners and managers. Source: Barber *et al.* (2014).

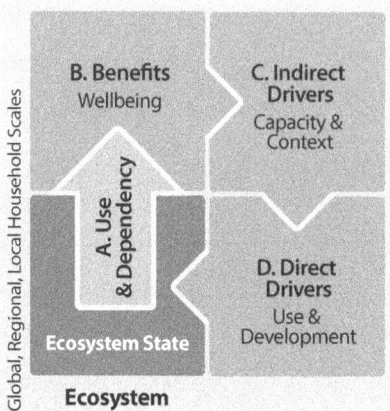

Fig. 3.4. Conceptual model for the SELTMP project, with the human dimension of the GBR represented by four components (in light blue). Source: Marshall *et al.* (2015).

involving Indigenous peoples, local communities and scientists, and subsequent email communication to refine and revise drafts (Lyver *et al.* 2015). In the Wet Tropics cluster case example, NRM practitioners and scientists co-produced infographics, thematic fact sheets and workshop presentations, and engaged in practitioner review of science products. For example, producing a communication product that showed the timeline of science discovery of climate change was identified by the NRM practitioners as vital for stakeholder communication. Several ways to present this were discussed between the science team collecting the relevant data, graphic designers, communications specialists and the NRM practitioners, followed by final review by the Australian Government funding agency. The resulting visual and written depiction is shown in Fig. 3.5.

Material domain capacities

The material domain concerns the tangible human resources, financial and structural capacities to sustain relationships between actors at different scales (Wyborn 2015). All the case studies reported sufficient resources to underpin the material capacities they needed, although some were clearly larger in scope and therefore required more resources, particularly to provide for necessary social interactions. For example, the Wet Tropics NRM Cluster Climate Adaptation project mobilised the concept of resources for brokerage, establishing a brokering hub and employing a knowledge broker (Bohnet *et al.* 2013). The global case examples required sufficient resources for global face-to-face dialogues. Local case examples required resources for camping trips onto country. In the GBR, substantial resources were required to support 11 working groups. In working with Indigenous peoples, there is a requirement to work at the pace and rhythm of customary practices, which means moving at a slower pace than in many other projects. Time becomes a vital resource.

Cross-scale linkages are recognised as one of the key co-productive capacities in the material domain (Wyborn 2015). In all the case examples, actors were engaged at local, regional, national and global domains. For example, the Traditional Owners involved in the fine-scale case study on Biocultural Mapping were also members of the sea-country forum, and linked with the national Indigenous Advisory Committee to the Minister for Environment. The scientists were linked with global initiatives in integration science through the STEPS Centre (Social, Technological and Environmental Pathways to

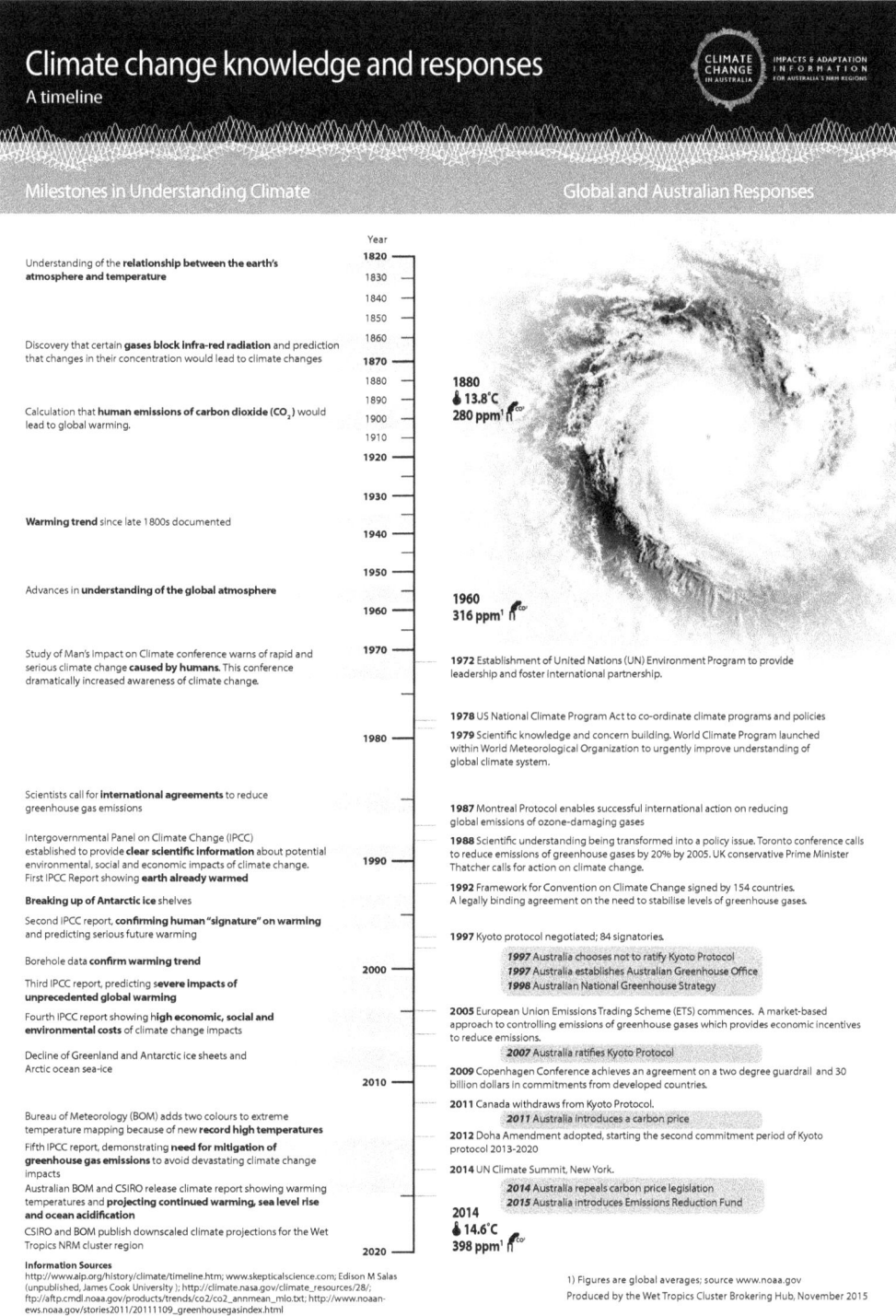

Fig. 3.5. Co-produced infographic developed for NRM stakeholder communication to show the accumulation of scientific knowledge about climate change over more than 130 years, together with global and Australian policy responses.

Sustainability). Collaboration between multiple actors enables cross-scale linkages and provides opportunity for mobilisation of institutions with scale-dependent comparative advantage (Hill *et al.* 2015a)

Transformative path-generation

Changes in ethics, values, beliefs, relationships, institutions, behaviours and actions which lead to a profound and fundamental change in human–environment interactions, characterise transformational pathways to sustainability (Westley *et al.* 2013). Many of these changes were identified in the case examples and are described below.

Changes in ethics: making the invisible visible was a theme in several cases. In the global pollination assessment, Indigenous peoples and local communities' priorities have been given a direct voice in a global assessment for the first time. This inclusion produced both new understanding of the relevance of their practices to the contemporary problem of pollinator decline, and a shift towards recognition of non-western knowledge systems, a key ethical principle for IPBES Conceptual (Díaz *et al.* 2015). Resources have been directed to local peoples to undertake work in their own communities on pollination. The Australian Indigenous Biocultural Knowledge website (www.aibk.info) resulted in a powerful visual representation of documented Indigenous peoples and their knowledge, as well as providing for online interactivity (Fig. 3.6).

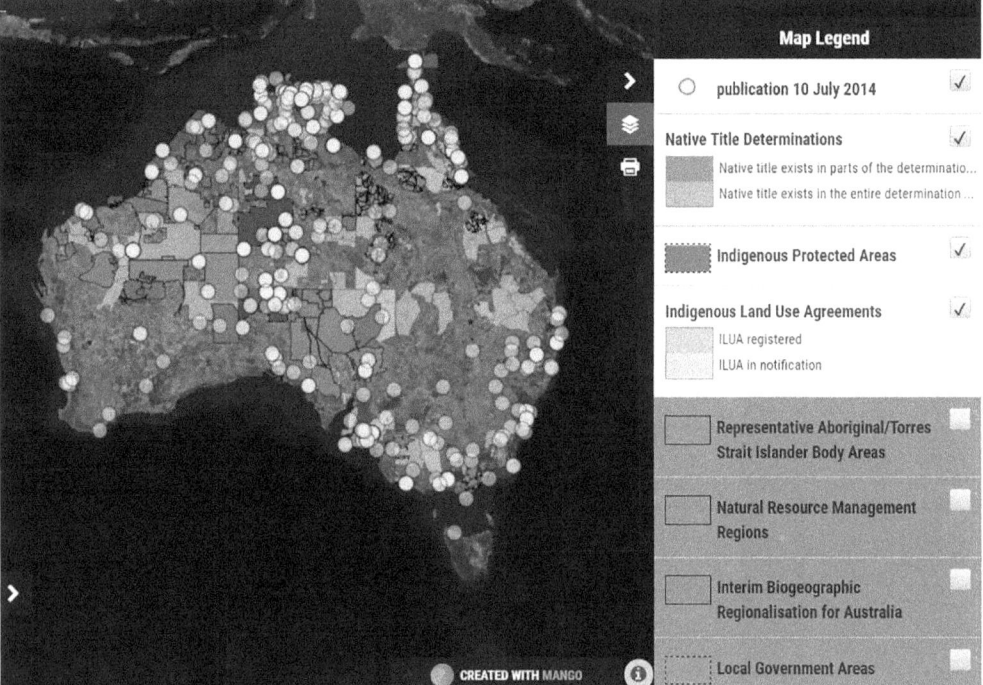

Fig. 3.6. Interactive map which allows users to visualise where Australian Indigenous Biocultural Knowledge has been documented. Source: Pert *et al.* 2015

Changes in beliefs: evidence for empowerment emerged in the case examples, a change in peoples' beliefs about their (and others') ability to bring about change. For example, an Ogiek person, a member of a highly marginalised community in Kenya which is taking the government to court over a human-rights abuse, was empowered and given a voice in a global pollination assessment, resulting in access to new networks and resources to support efforts to protect the community's human rights. The SELTMP project identified people wanting to do more to protect the GBR and some wanting to be more involved in management. New initiatives are underway to provide greater opportunities for community engagement, through the Reef 2050 Long-term Sustainability Plan.

Changes in values: shifting towards greater recognition of Indigenous peoples and their roles was a context, an outcome and an ongoing need that emerged through the case examples. Until the 1990s, Australian environmental policy, legislation and ideas largely focused on ameliorating the impacts of industrial agriculture, tourism, mining and forestry on the unique biota, climate and landscapes (Mercer 1991). Land tenure regimes historically reflected and resulted in colonial dispossession.

However, several decades of native title recognition and Indigenous land restitution initiatives have seen a fundamental recasting of the Australian landscapes. Much of Australia's high-value urban, agriculture and industrial landscapes have remained out of the reach of these initiatives, but by 2013 more than half of the Australian continent had Indigenous rights and interests recognised, including 16% as tenure (Hill *et al.* 2013b). By 2015, more than 43% of Australia's National Reserve System for protecting nature was made up of Indigenous Protected Areas, covering over 60 million hectares. Increasing the formal roles of Indigenous peoples in environmental management was recognised as a stand-out trend in Australia's State of the Environment during the first decade of the 21st century (State of the Environment Committee 2011). Several of our case examples were established to give recognition and priority to Indigenous knowledge (e.g. Barber *et al.* 2015); in others, the recognition was an outcome of the real-world encounter between scientists, practitioners and policy actors (e.g. Hill *et al.* 2015b).

Globally, similar shifts are underway, with the proportion of forests in developing countries recognised as under the ownership of or designated for Indigenous peoples and local communities growing from 22% to 30% over the last decade (RRI 2015). The UN Declaration on the Rights of Indigenous Peoples has been adopted by 146 countries, and the Intergovernmental Platform on Biodiversity and Ecosystem Services is among several UN-mandated organisations recognising the need for new approaches to research with Indigenous and local knowledge.

Despite these changes in recognition of Indigenous rights, key gaps remain in the capability to integrate Indigenous knowledge. The GBR Social and Economic Long-term Monitoring project did not capture any data relevant to Indigenous use, dependency and well-being, despite its recognition of the need for effective monitoring of the many roles of Indigenous peoples in the GBR, and its efforts to ensure inclusion (Nursey-Bray and Rist 2009; Nursey-Bray and Jacobson 2014; Addison *et al.* 2016). The analysis of the global biodiversity targets highlighted how the targets related to Indigenous knowledge and Indigenous rights required no change to discriminatory policies of nation-states, thereby worsening current inequities while appearing to support change (Hill *et al.* 2015c).

Changes in institutions and relationships: new collaborative institutions were initiated in several of the case examples. In the Wet Tropics NRM case example, the brokering hub that was established to support knowledge exchange is viewed by the project partners as a valuable institutional arrangement, which they are seeking to extend beyond the initial three-year life of the project. The brokering model empowers the practitioners to challenge scientists to be more practical and specific, leading to science products that are more capable of supporting change in on-ground actions.

The NICC project resulted in provisions in Australian legislation and policy for Indigenous peoples' roles in carbon abatement. The project recognised that climate change mitigation opportunities need to be treated as an integral element of sustainable development:

> *Strategies for mitigation and adaptation must be holistic, taking into account not only the ecological dimensions of climate change, but also the social impacts, human rights, equity and environmental justice. Indigenous peoples, who have the smallest ecological footprints, should not be asked to carry the heavier burden of adjusting to climate change (UN Permanent Forum on Indigenous Issues 2008).*

Primary and secondary statutory recognition of Indigenous peoples' rights and interests in land and water was critical to the NICC project. These rights and interests are important drivers in preserving space for Indigenous land management and appropriate payment for ecosystem services in Australia (Gerrard 2008). These legal frameworks continue to be decoded through judicial decisions and government policy reform, and have different implications for Indigenous communities across Australia who wish to establish climate change mitigation projects on their land (Gerrard 2012). Indigenous organisations have difficulty adapting their existing operations and activities to pursue carbon mitigation opportunities, with many organisations reporting that they did not have access to the appropriate information and resources that would enable them to participate in existing carbon market opportunities (Robinson *et al.* 2014a). Ongoing auditing and other operational requirements added a prohibitive administrative and financial burden to resource-poor Indigenous organisations.

Changes in behaviours: the use of language. In the Reef Catchments case example, changes in language, in how keywords were used and understood, reflected a significant shift in both behaviours and relationships. The term 'Traditional Owner' was greatly enriched by mutual experiences that revealed the history of connection between the Traditional Owners, their culture, their cultural sites and their country. The experiences also revealed the need to address the impacts of that history, evident in ongoing marginalisation of people, and barriers to accessing their traditional land. In the NICC example, the re-conceptualising of how carbon PES schemes could generate benefits for Indigenous people, and how such benefits might be assessed, fundamentally changed the framing and understanding of climate change as an opportunity as well as a threat.

Discussion and conclusion

Solving sustainability problems related to land and water often involves multiple stakeholders with different knowledge bases and different goals. Effective science to underpin solution-generation demands the reconciliation of these differences, and opens up critical relationships between sustainability science and action that emphasises new roles for scientists, community and civil society. Case examples of integration science outlined in this chapter highlight changes to ethics, beliefs, values, institutions, relationships and behaviours – the types of changes identified as necessary to the profound and fundamental change in human–environment interactions that characterise transformational pathways to sustainability (Fig. 3.7). Transformative pathways open up new understandings of environmental problems and alternative and agreed courses of action, based on a broad set of

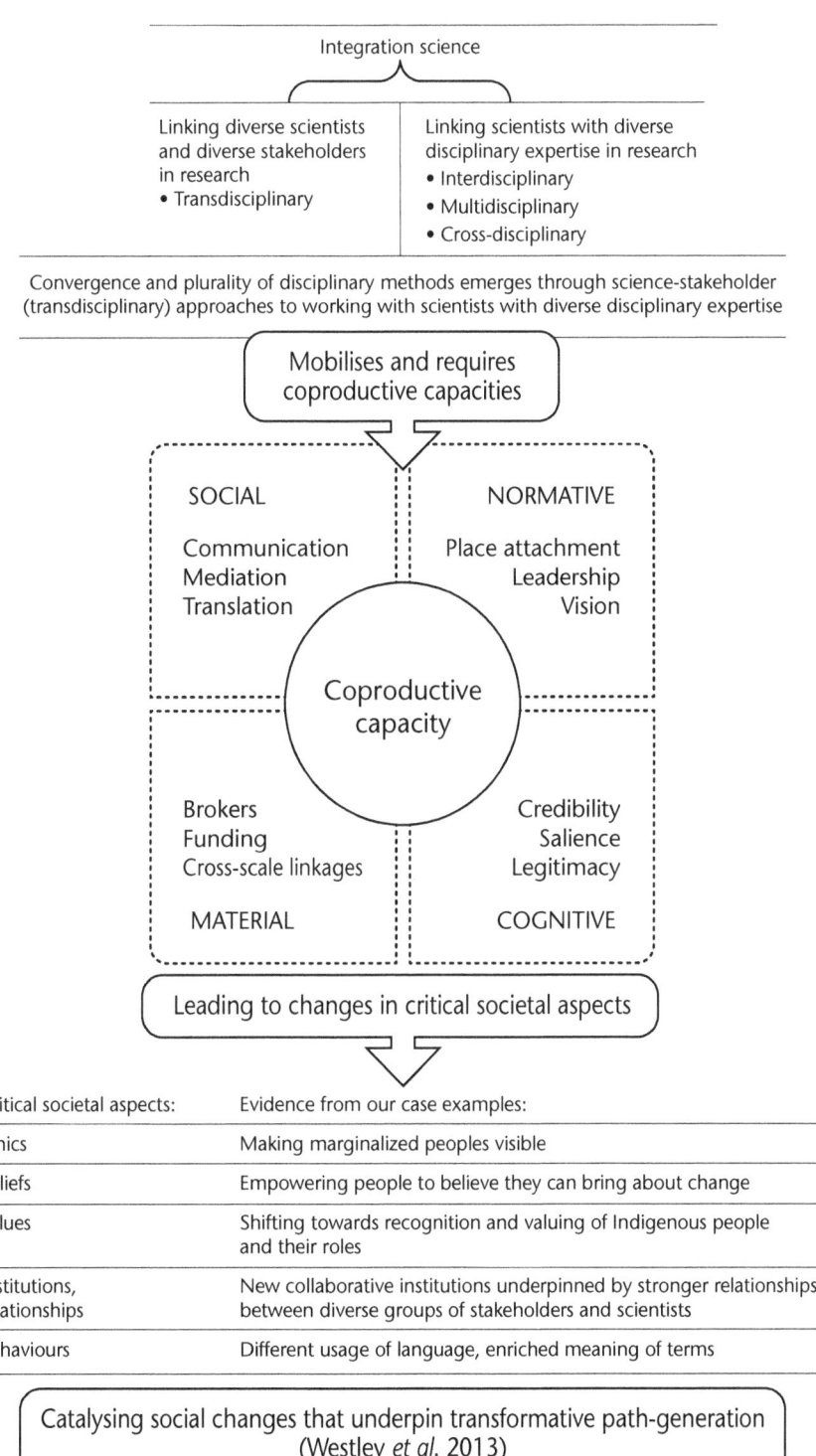

Fig. 3.7. Integration science, co-productive capacities and the changes that underpin transformative path-generation.

knowledge contribution. They inevitably challenge the powerful, who are vested in maintaining current conditions (O'Brien 2012).

Understanding the histories, contexts and relationships from which the current relations between science and governance have emerged is central to enabling mobilisation of capacities for particular environmental problems (Wyborn 2015). In Australia, as several other countries, a key context for land and water sustainability involves Indigenous peoples, country, knowledge and rights. Political and legal recognition of the role of Indigenous Australians in managing Australia's environment has matured from imported European perceptions of Australia as an empty Indigenous 'wilderness' to legal recognition of native title rights and Indigenous land management systems (Hill et al. 2013b). The case examples involving Indigenous peoples provide evidence that integration science has the potential to shift towards contexts that will enable recognition of Indigenous peoples' roles and responsibilities; however, achieving this potential requires specific techniques and approaches. The ongoing marginalisation of Indigenous voices is also evident from the case examples.

Indigenous scholars have persistently argued that this marginalisation results from continuation of research practices that delegitimise and disempower Indigenous people (Smith 1999). The marginalisation exacerbates the consequences of separation from their knowledge, culture, lands and families during colonisation, and prevents knowledge and language (among other things) being transferred to younger generations (Martin and Mirraboopa 2003; Louis 2007). Existing research practices are limited in their capacity to address a diversity of worldviews, to meet requirements for protection of cultural and intellectual rights and to recognise existing Indigenous governance systems (Liddle 2001; Hankins and Ross 2008; Simonds and Christopher 2013). Unequal power relationships further entrench the exclusion of Indigenous voices in research (Hankins and Ross 2008). Langton (1998) highlights how misrepresentations, tropes and asymmetric power relationships impeded for decades the effective application of Aboriginal knowledge of fire management – knowledge that has now been recognised as vital to solving global problems of excessive greenhouse gas emissions and biodiversity loss in tropical savannas (Russell-Smith et al. 2013).

Indigenous scientists are now establishing Indigenous research methods that enable their perspectives, voices and knowledge systems to be equitably reflected in land and water science (Johnson et al. 2016; Kealiikanakaoleohaililani and Giardina 2016; Whyte et al. 2016). Indigenous research reflects lived and shared cultural experiences, histories and relational responsibility, and accountability by the researcher to the Indigenous community and the Indigenous scientist's own community. These approaches present a vigorous alternative means to enable path-generation that takes into account Indigenous rights and interests in land and water sustainability. Further exploration and adoption of the Indigenous research methods of Indigenous scientists may better enable their perspectives, voices and knowledge systems to be equitably respected and reflected in land and water science.

The case examples presented here show that integration science can mobilise co-productive capacities across normative, social cognitive and material domains. The values that underpin the normative domain may be contested and difficult to negotiate, but they are critical. The social domain requires scientific attention to the design of effective and equitable governance systems surrounding sustainability issues, including the governance of research projects and partnerships. This can be challenging in contexts where the type of knowledge deemed credible may be contested, and the process through which knowledge is legitimised may need careful negotiation. The case studies that draw on integration science activities with Indigenous people highlight the important role of scientists as brokers. Scientific constructs (e.g. models, graphs and maps) and scientific communications (e.g. papers and research forums) can help to negotiate between boundaries of knowledge generation, sharing and translation practice (Davies *et al.* 2013). Participatory modelling and negotiation of co-produced knowledge outputs are key tools for mobilising capacities in the cognitive domain. Mobilising the material domain capacities highlights the need for sufficient resources, particularly time, to provide for the necessary social interaction between actors.

There is no doubt that integration science approaches challenge conventional methods and evaluations of research impact. These research efforts and partnerships require that different values, expectations, knowledge systems and reward paradigms be taken into account. Assessment of the creation of impact through integration science may need to consider multiple time-lags between intervention and response, different and interacting drivers of change, and the need to account for less tangible outcomes (e.g. social equity and empowerment). Scientists taking on integration science also recognise they are part of rather than separate from transformative pathways towards sustainability. This demands a research paradigm that views science as one part of a network of knowledge partnerships, rather than the science communication model that creates a research provider/research user dynamic (Michaels 2009). Scientists need to have inclusive and co-reflective brokering skills to navigate the protocols involved in bridging different knowledge systems, and pragmatic and context-relevant brokering skills to handle the politics involved at the interface between knowledge exchange, inclusion and co-production. In all the case examples, the capacity of scientists, communities, policy-makers and on-ground practitioners to co-invest in new knowledge was critical to generating the pathways towards sustainability.

Acknowledgements

We would like to acknowledge and thank the funding and in-kind support from numerous agencies that made this research possible, including the Intergovernmental Platform on Biodiversity and Ecosystem Services, UNESCO, the Smithsonian Institute, the US Department of Agriculture, the UN Food and Agriculture Organization, CSIRO's former Biodiversity Theme, the Australian government's National Environmental Research Program, the James Hutton Institute, the Scottish government's Rural Affairs and Environment Portfolio Strategic Research Programme, the Australian Centre for Ecological Analysis and Synthesis, the National Collaborative Research Infrastructure Strategy and Education Infrastructure Fund, the Great Barrier Reef Marine Park Authority, James Cook University, the Australian government's Natural Resource Management Climate Change Impacts and Adaptation Research Program, Terrain NRM, Reef Catchments NRM, Cape York NRM, Torres Strait Regional Authority, Kowanyama Land and Natural Resource Management Office, CSIRO's former Water for a Healthy Country Flagship and Climate Adaptation Flagship.

References

Addison P, Walshe T, Sweatman H, Jonker M, Anthony K, MacNeil A, Thompson A, Logan M (2016) *Towards an Integrated Monitoring Program: Identifying Indicators and Existing Monitoring Programs to Effectively Evaluate the Long Term Sustainability Plan*. Report to the National Environmental Science Program. Reef and Rainforest Research Centre,

Cairns. <http://nesptropical.edu.au/wp-content/uploads/2016/01/NESP-TWQ-3.8-FINAL-REPORT.pdf>.

Adger WN, Barnett J, Brown K, Marshall N, O'Brien K (2013) Cultural dimensions of climate change impacts and adaptation. *Nature Climate Change* **3**(2), 112–117. doi:10.1038/nclimate1666

Barber M, Jackson S, Shellberg J, Sinnamon V (2014) Working knowledge: characterising collective Indigenous, scientific, and local knowledge about the ecology, hydrology and geomorphology of Oriners Station, Cape York Peninsula, Australia. *Rangeland Journal* **36**(1), 53–66. doi:10.1071/RJ13083

Barber M, Jackson S, Dambacher J, Finn M (2015) The persistence of subsistence: qualitative social-ecological modeling of Indigenous aquatic hunting and gathering in tropical Australia. *Ecology and Society* **20**(1), 60. doi:10.5751/ES-07244-200160

Bohnet IC, Hill R, Turton SM, Bell R, Hilbert DW, Hinchley D, Pressey B, Rainbird J, Standley P-M, Cvitanovic C, Crowley GM, Curnock M, Dale A, Lyons P, Moran C, Pert PL (2013) Supporting Regional Natural Resource Management (NRM) organisations to update their NRM plans for adaptation to climate change. In *Adapting to Change: The Multiple Roles of Modelling.* 20th International Congress on Modelling and Simulation (MODSIM), Adelaide.

Brito L, Stafford Smith M (2012) *State of the Planet Declaration.* Key Messages from the Planet under Pressure Conference, London. 26–29 March. <http://www.planetunderpressure2012.net/pdf/state_of_planet_declaration.pdf>.

Brown VA, Harris JA, Russell JY (Eds) (2010) *Tackling Wicked Problems: Through the Transdisciplinary Imagination.* Earthscan, London.

Campbell CA, Lefroy EC, Caddy-Retalic A, Bax N, Doherty PF, Douglas MM, Johnson D, Possingham HP, Specht A, Tarte D, West J (2015) Designing environmental research for impact. *Science of the Total Environment* **534**, 4–13. doi:10.1016/ j.scitotenv.2014.11.089.

Câmpeanu CN, Fazey I (2014) Adaptation and pathways of change and response: a case study from eastern Europe. *Global Environmental Change* **28**, 351–367. doi:10.1016/j.gloenvcha.2014.04.010

Cash DW, Clark WC, Alcock F, Dickson NM, Eckley N, Guston DH, Jager J, Mitchell RB (2003) Knowledge systems for sustainable development. *Proceedings of the National Academy of Sciences of the United States of America* **100**, 8086–8091. doi:10.1073/pnas.1231332100

Costanza R, de Groot R, Sutton P, van der Ploeg S, Anderson SJ, Kubiszewski I, Farber S, Turner RK (2014) Changes in the global value of ecosystem services. *Global Environmental Change* **26**, 152–158. doi:10.1016/j.gloenvcha.2014.04.002

Crotty M (1998) *The Foundations of Social Research.* Allen and Unwin, Sydney.

Davies J, Hill R, Walsh FJ, Sandford M, Smyth D, Holmes MC (2013) Innovation in management plans for Community Conserved Areas: experiences from Australian Indigenous Protected Areas. *Ecology and Society* **18**(2), 14. doi:10.5751/ES-05404-180214

Davies JM, Beggs PJ, Medek DE, Newnham RM, Erbas B, Thibaudon M, Katelaris CH, Haberle SG, Newbigin EJ, Huete AR (2015) Trans-disciplinary research in synthesis of grass pollen aerobiology and its importance for respiratory health in Australasia. *Science of the Total Environment* **534**, 85–96. doi:10.1016/j.scitotenv.2015.04.001

Descola P (2014) Modes of being and forms of predication. *Journal of Ethnographic Theory* **4**(1), 271–280. doi:10.14318/hau4.1.012

Díaz S, Demissew S, Carabias J, Joly C, Lonsdale M, Ash N, Larigauderie A, Adhikari JR, Arico S, Báldi A, Bartuska A, Baste IA, Bilgin A, Brondizio E, Chan KMA, Figueroa VE, Duraiappah A, Fischer M, Hill R, Koetz T, Leadley P, Lyver P, Mace GM, Martin-Lopez B, Okumura M, Pacheco D, Pascual U, Pérez ES, Reyers B, Roth E, Saito O, Scholes RJ, Sharma

N, Tallis H, Thaman R, Watson R, Yahara T, Hamid ZA, Akosim C, Al-Hafedh Y, Allahverdiyev R, Amankwah E, Asah ST, Asfaw Z, Bartus G, Brooks LA, Caillaux J, Dalle G, Darnaedi D, Driver A, Erpul G, Escobar-Eyzaguirre P, Failler P, Fouda AMM, Fu B, Gundimeda H, Hashimoto S, Homer F, Lavorel S, Lichtenstein G, Mala WA, Mandivenyi W, Matczak P, Mbizvo C, Mehrdadi M, Metzger JP, Mikissa JB, Moller H, Mooney HA, Mumby P, Nagendra H, Nesshover C, Oteng-Yeboah AA, Pataki G, Roué M, Rubis J, Schultz M, Smith P, Sumaila R, Takeuchi K, Thomas S, Verma M, Yeo-Chang Y, Zlatanova D (2015) The IPBES Conceptual Framework: connecting nature and people. *Current Opinion in Environmental Sustainability* **14**, 1–16. doi:10.1016/j.cosust.2014.11.002

Djelic ML, Quack S (2007) Overcoming path dependency: path generation in open systems. *Theory and Society* **36**(2), 161–186. doi:10.1007/s11186-007-9026-0

Ens EJ, Pert P, Clarke PA, Budden M, Clubb L, Doran B, Douras C, Gaikwad J, Gott B, Leonard S, Locke J, Packer J, Turpin G, Wason S (2015) Indigenous biocultural knowledge in ecosystem science and management: review and insight from Australia. *Biological Conservation* **181**, 133–149. doi:10.1016/j.biocon.2014.11.008

Fazey I, Bunse L, Msika J, Pinke M, Preedy K, Evely AC, Lambert E, Hastings E, Morris S, Reed MS (2014) Evaluating knowledge exchange in interdisciplinary and multi-stakeholder research. *Global Environmental Change* **25**, 204–220. doi:10.1016/j.gloenvcha.2013.12.012

Geels FW (2010) Ontologies, socio-technical transitions (to sustainability), and the multi-level perspective. *Research Policy* **39**, 495–510. doi:10.1016/j.respol.2010.01.022

Gelcich S, Hughes TP, Olsson P, Folke C, Defeo O, Fernández M, Foale S, Gunderson LH, Rodríguez-Sickert C, Scheffer M, Steneck RS, Castilla JC (2010) Navigating transformations in governance of Chilean marine coastal resources. *Proceedings of the National Academy of Sciences of the United States of America* **107**, 16794–16799. doi:10.1073/pnas.1012021107

Gerrard E (2008) Climate change and human rights: issues and opportunities for Indigenous peoples. *UNSW Law Journal* **31**, 941.

Gerrard E (2012) Towards a carbon constrained future: climate change, emissions trading and Indigenous peoples' rights in Australia. In *Native Title and Ecology*. (Ed. J Weir) pp. 135–174. ANU Press, Canberra.

Goldberg J, Marshall N, Birtles A, Case P, Bohensky E, Curnock M, Gooch M, Parry-Husbands H, Pert P, Tobin R, Villani C, Visperas B (2016) Climate change, the Great Barrier Reef and the response of Australians. *Palgrave Communications* **2**, 15046. doi:10.1057/palcomms.2015.46

Hankins DL, Ross J (2008) Research on native terms: navigation and participation issues for native scholars in community research. In *Partnerships for Empowerment: Participatory Research for Community-based Natural Resource Management*. (Eds C Wilmsen, W Elmendorf, L Fisher, J Ross, B Sarathy, G Wells) pp. 239–257. Earthscan, London.

Hendriks CM (2009) Deliberative governance in the context of power. *Policy and Society* **28**, 173–184. doi:10.1016/j.polsoc.2009.08.004

Hill R, Grant C, George M, Robinson CJ, Jackson S, Abel N (2012) A typology of Indigenous engagement in Australian environmental management: implications for knowledge integration and social-ecological system sustainability. *Ecology and Society* **17**, 23. doi:10.5751/ES-04587-170123

Hill R, Halamish E, Gordon IJ, Clark M (2013a) The maturation of biodiversity as a global social-ecological issue and implications for future biodiversity science and policy. *Futures* **46**, 41–49. doi:10.1016/j.futures.2012.10.002

Hill R, Pert PL, Davies J, Robinson CJ, Walsh F, Falco-Mammone F (2013b) *Indigenous Land Management in Australia. Diversity, Scope, Extent, Success Factors and Barriers*. CSIRO

Ecosystem Sciences, Cairns. <http://www.daff.gov.au/__data/assets/pdf_file/0010/2297116/ilm-report.pdf>.

Hill R, Pert PL, Barber M, Robinson CJ, Willimas V, Jenkins S (2015a) *Transdisciplinary and Interdisciplinary Integration Science (TIIS): Strategy Development*. CSIRO Land and Water Strategic Project Report. CSIRO, Cairns.

Hill R, Pert PL, Lyons I, Turton SM, Moran C (Eds) (2015b) *Uptake of Climate Change Adaptation Pathways and Opportunities into Wet Tropics Cluster NRM Plans: System Building in Progress*. CSIRO Land and Water Flagship and James Cook University, Cairns. <https://terranova.org.au/repository/wet-tropics-nrm-collection/uptake-of-climate-change-adptation-pathways-and-opportunities-into-wet-tropics-cluster-nrm-plans-system-building-in-progress>.

Hill R, Dyer GA, Lozada-Ellison LM, Gimona A, Martin-Ortega J, Munoz-Rojas J, Gordon IJ (2015c) A social-ecological systems analysis of impediments to delivery of the Aichi 2020 iargets and potentially more effective pathways to the conservation of biodiversity. *Global Environmental Change* **34**, 22–34. doi:10.1016/j.gloenvcha.2015.04.005

Hill R, Kwapong PK, Nates-Parra G, Breslow S, Buchori D, Howlett B, LeBuhn G, Maués MM, Quezada-Euán JJ, Saeed S (2016) Biocultural diversity, pollinators and their socio-cultural values. Report, Deliverable 3a: *Thematic Assessment on Pollinators, Pollination and Food Production*, ch. 5. Intergovernmental Platform on Biodiversity and Ecosystem Services. Bonn, Germany.

Jacobs K, Lebel L, Buizer J, Addams L, Matson P, McCullough E, Garden P, Saliba G, Finan T (2010) Linking knowledge with action in the pursuit of sustainable water-resources management. *Proceedings of the National Academy of Sciences* **113**, 4591–4596. doi:10.1073/pnas.0813125107

Johnson JT, Howitt R, Cajete G, Berkes F, Louis RP, Kliskey A (2016) Weaving Indigenous and sustainability sciences to diversify our methods. *Sustainability Science* **11**, 1–11. doi:10.1007/s11625-015-0349-x

Kealiikanakaoleohaililani K, Giardina CP (2016) Embracing the sacred: an Indigenous framework for tomorrow's sustainability science. *Sustainability Science* **11**, 57–67. doi:10.1007/s11625-015-0343-3

Khagram S, Nicholas KA, Bever DM, Warren J, Richards EH, Oleson K, Kitzes J, Katz R, Hwany R, Goldman R, Funk J, Brauman KA (2010) Thinking about knowing: conceptual foundations for interdisciplinary environmental research. *Environmental Conservation* **37**, 388–397. doi:10.1017/S0376892910000809

Kinzig AP (2001) Bridging disciplinary divides to address environmental and intellectual challenges. *Ecosystems* **4**, 709–715. doi:10.1007/s10021-001-0039-7

Langton M (1998) *Burning Questions: Emerging Environmental Issues for Indigenous Peoples in Northern Australia*. Centre for Indigenous Natural and Cultural Resource Management, Northern Territory University, Darwin.

Leach M, Scoones I, Stirling A (2010) *Dynamic Sustainabilities: Technology, Environment, Social Justice*. Earthscan/Economic and Social Research Council, London.

Leach M, Rockstrom J, Raskin P, Scoones I, Stirling AC, Smith A, Thompson J, Millstone E, Ely A, Arond E, Folke C, Olsson P (2012) Transforming innovation for sustainability. *Ecology and Society* **17**, 11. doi:10.5751/ES-04933-170211

Leach M, Raworth K, Rockström J (2013) Between social and planetary boundaries: navigating pathways in the safe and just space for humanity. In *World Social Science Report. Changing Global Environments*. (Eds ISSC and UNESCO) pp. 84–89. OECD Publishing/UNESCO Publishing, Paris.

Liddle L (2001) Bridging the communication gap: transferring information between scientists and Aboriginal land managers. In *Working on Country: Contemporary Indigenous Management of Australia's Lands and Coastal Regions*. (Eds R Baker, J Davies and E Young) pp. 147–155. Oxford University Press, Melbourne.

López B, Miro FL, López A, López EEG (2015) Guna people of Panama: Indigenous and local knowledge about pollination and pollinators associated with food production. In *Indigenous and Local Knowledge about Pollination and Pollinators associated with Food Production: Outcomes from the Global Dialogue Workshop*. Panama. 1–5 December 2014. (Eds P Lyver, E Perez, M Carneiro da Cunha and M Roué) pp. 38–45. UNESCO, Paris.

Louis RP (2007) Can you hear us now? Voices from the margin: using Indigenous methodologies in geographic research. *Geographical Research* **45**, 130–139. doi:10.1111/j.1745-5871.2007.00443.x

Lynch AJJ, Thackway R, Specht A, Beggs PJ, Brisbane S, Burns EL, Byrne M, Capon SJ, Casanova MT, Clarke PA, Davies JM, Dovers S, Dwyer RG, Ens E, Fisher DO, Flanigan M, Garnier E, Guru SM, Kilminster K, Locke J, MacNally R, McMahon KM, Mitchell PJ, Pierson JC, Rodgers EM, Russell-Smith J, Udy J, Waycott M (2015) Transdisciplinary synthesis for ecosystem science, policy and management: the Australian experience. *Science of the Total Environment* **534**, 173–184. doi:10.1016/j.scitotenv.2015.04.100

Lyver P, Perez E, Carneiro da Cunha M, Roué M (Eds) (2015) *Indigenous and Local Knowledge about Pollination and Pollinators Associated with Food Production: Outcomes from the Global Dialogue Workshop*. Panama. 1–5 December 2014. <http://www.unesco.org/new/fileadmin/MULTIMEDIA/HQ/SC/pdf/IPBES_Pollination-Pollinators_Panama_Workshop.pdf>.

Marshall NA, Bohensky E, Curnock M, Goldberg J, Gooch M, Nicrota B, Pert PL, Scherl L, Stone-Jovicich S, Tobin RC (2015) *The Social and Economic Long-term Monitoring Program for the Great Barrier Reef 2014 Final Report*. Report to the National Environmental Research Program. Reef and Rainforest Research Centre, Cairns.

Martin K, Mirraboopa B (2003) Ways of knowing, being and doing: a theoretical framework and methods for Indigenous and indigenist re-search. *Journal of Australian Studies* **27**, 203–214. doi:10.1080/14443050309387838

Mercer D (1991) *A Question of Balance: Natural Resources Conflict Issues in Australia*. Federation Press, Sydney.

Michaels S (2009) Matching knowledge brokering strategies to environmental policy problems and settings. *Environmental Science and Policy* **12**, 994–1011. doi:10.1016/j.envsci.2009.05.002

NRC (National Research Council) (2014) *Convergence: Facilitating Transdisciplinary Integration of Life Sciences, Physical Sciences, Engineering, and Beyond*. National Academies Press, Washington DC.

Nursey-Bray M, Jacobson C (2014) 'Which way?': the contribution of Indigenous marine governance. *Australian Journal of Maritime and Ocean Affairs* **6**, 27–40. doi:10.1080/18366503.2014.888136

Nursey-Bray M, Rist P (2009) Co-management and protected area management: achieving effective management of a contested site, lessons from the Great Barrier Reef World Heritage Area (GBRWHA). *Marine Policy* **33**, 118–127. doi:10.1016/j.marpol.2008.05.002

O'Brien K (2012) Global environmental change II: from adaptation to deliberate transformation. *Progress in Human Geography* **36**, 667–676. doi:10.1177/0309132511425767

Olsson P, Folke C, Hughes TP (2008) Navigating the transition to ecosystem-based management of the Great Barrier Reef, Australia. *Proceedings of the National Academy of Sciences of the United States of America* **105**, 9489–9494. doi:10.1073/pnas.0706905105

Olsson P, Galaz V, Boonstra WJ (2014) Sustainability transformations: a resilience perspective. *Ecology and Society* 19(4), 1. doi:10.5751/ES-06799-190401

Pereira L, Karpouzoglou T, Doshi S, Frantzeskaki N (2015) Organising a safe space for navigating social-ecological transformations to sustainability. *International Journal of Environmental Research and Public Health* 12(6), 6027–6044. doi:10.3390/ijerph120606027

Pert PL, Ens EJ, Locke J, Clarke PA, Packer JM, Turpin G (2015) An online spatial database of Australian Indigenous biocultural knowledge for contemporary natural and cultural resource management. *Science of the Total Environment* 534, 110–121. doi:10.1016/j.scitotenv.2015.01.073

Podestá GP, Natenzon CE, Hidalgo C, Toranzo FR (2013) Interdisciplinary production of knowledge with participation of stakeholders: a case study of a collaborative project on climate variability, human decisions and agricultural ecosystems in the Argentine Pampas. *Environmental Science and Policy* 26, 40–48. doi:10.1016/j.envsci.2012.07.008

Potts SG, Imperatriz-Fonseca VL, Ngo HT, Biesmeijer JC, Breeze TD, Dicks LV, Garibaldi LA, Hill R, Settele J, Vanbergen AJ, Aizen MA, Cunningham SA, Eardley C, Freitas BM, Gallai N, Kevan PG, Kovács-Hostyánszki A, Kwapong PK, Li J, Li X, Martins DG, Nates-Parra G, Pettis JS, Viana BF (Eds) (2016) IPBES 2016 Summary for policymakers. In *Pollinators, Pollination and Food Production. Contribution of the Expert Group to the First Assessment Report (Deliverable 3a) of the Intergovernmental Platform on Biodiversity and Ecosystem Services.* IPBES Secretariat, Bonn.

Robinson CJ, Wallington T (2012) Boundary work: engaging knowledge systems in co-management of feral animals on Indigenous lands. *Ecology and Society* 17, 16. doi:10.5751/ES-04836-170216

Robinson CJ, Gerrard E, May T, Maclean K (2014a) Australia's Indigenous carbon economy: a national snapshot. *Geographical Research* 52, 123–132. doi:10.1111/1745-5871.12049

Robinson CJ, Taylor B, Vella K, Wallington T (2014b) Working knowledge for collaborative water planning in Australia's Wet Tropics region. *International Journal of Water Governance* 2, 43–60. doi:10.7564/13-IJWG4

Robinson CJ, James G, Whitehead PJ (2016a) Negotiating Indigenous benefits from payment for ecosystem service (PES) schemes. *Global Environmental Change* 38, 21–29. doi:10.1016/j.gloenvcha.2016.02.004

Robinson CJ, Renwick AR, May T, Gerrard E, Foley R, Battaglia M, Possingham H, Griggs D, Walker D (2016b) Indigenous benefits and carbon offset schemes: an Australian case study. *Environmental Science and Policy* 56, 129–134. doi:10.1016/j.envsci.2015.11.007

Rockström J, Steffen W, Noone K, Persson A, Chapin FS, Lambin EF, Lenton TM, Scheffer M, Folke C, Schellnhuber HJ, Nykvist B, de Wit CA, Hughes T, van der Leeuw S, Rodhe H, Sorlin S, Snyder PK, Costanza R, Svedin U, Falkenmark M, Karlberg L, Corell RW, Fabry VJ, Hansen J, Walker B, Liverman D, Richardson K, Crutzen P, Foley JA (2009) A safe operating space for humanity. *Nature* 461, 472–475. doi:10.1038/461472a

Rosenfield PL (1992) Special Issue. Building research capacity for health social sciences in developing countries: the potential of transdisciplinary research for sustaining and extending linkages between the health and social sciences. *Social Science and Medicine* 35, 1343–1357. doi:10.1016/0277-9536(92)90038-R

RRI (2015) *Who Owns the World's Land? A Global Baseline of Formally Recognized Indigenous and Community Land Rights*. Rights and Resources Initiative, Washington DC.

Russell-Smith J, Cook GD, Cooke PM, Edwards AC, Lendrum M, Meyer CP, Whitehead PJ (2013) Managing fire regimes in north Australian savannas: applying Aboriginal approaches to contemporary global problems. *Frontiers in Ecology and the Environment* 11, e55–e63. doi:10.1890/120251

Secretariat of the Convention on Biological Diversity (2011) *Strategic Plan for Biodiversity 2011–2020 and the Aichi Targets: Living in Harmony with Nature*. Secretariat of the Convention on Biological Diversity Montreal. <http://www.cbd.int/doc/strategic-plan/2011-2020/Aichi-Targets-EN.pdf>.

Sen A (2005) Human rights and capabilities. *Journal of Human Development* **6**, 151–166. doi:10.1080/14649880500120491

Sen A (2013) The ends and means of sustainability. *Journal of Human Development and Capabilities* **14**, 6–20. doi:10.1080/19452829.2012.747492

Simonds VW, Christopher S (2013) Adapting western research methods to Indigenous ways of knowing. *American Journal of Public Health* **103**, 2185–2192. doi:10.2105/AJPH.2012.301157

Smith LT (1999) *Decolonizing Methodologies: Research and Indigenous Peoples*. Zed Books/University of Otago Press, London and Dunedin.

Specht A, Guru S, Houghton L, Keniger L, Driver P, Ritchie EG, Lai K, Treloar A (2015) Data management challenges in analysis and synthesis in the ecosystem sciences. *Science of the Total Environment* **534**, 144–158. doi:10.1016/j.scitotenv.2015.03.092.

Stahl C, Cimorelli A (2013) A demonstration of the necessity and feasibility of using a clumsy decision analytic approach on wicked environmental problems. *Integrated Environmental Assessment and Management* **9**, 17–30. doi:10.1002/ieam.1356

State of the Environment Committee (2011) *Australia: State of the Environment 2011*. Independent report to the Federal Minister for Sustainability, Environment, Water, Population and Communities. DSEWPaC, Canberra.

Stokols D, Hall KL, Taylor BK, Moser RP (2008) The science of team science: overview of the field and introduction to the supplement. *American Journal of Preventive Medicine* **35**, S77–S89. doi:10.1016/j.amepre.2008.05.002

Tress G, Tress B, Fry G (2005) Clarifying integrative research concepts in landscape ecology. *Landscape Ecology* **20**, 479–493. doi:10.1007/s10980-004-3290-4

UN Permanent Forum on Indigenous Issues (2008) *Report on the Seventh Session (21 April–2 May 2008)*. Economic and Social Council Official Records Supplement No. 23, E/2008/43, E/C.19/2008/13. United Nations, New York.

van Kerkhoff LE, Lebel L (2015) Coproductive capacities: rethinking science–governance relations in a diverse world. *Ecology and Society* **20**(1), 14. doi:10.5751/ES-07188-200114

Van Opstal M, Huge J (2013) Knowledge for sustainable development: a worldviews perspective. *Environment, Development and Sustainability* **15**, 687–709. doi:10.1007/s10668-012-9401-5

Walsh FJ, Dobson PV, Douglas JC (2013) Anpernirrentye: a framework for enhanced application of Indigenous ecological knowledge in natural resource management. *Ecology and Society* **18**(3), 18. doi:10.5751/ES-05501-180318

Westley FR, Tjornbo O, Schultz L, Olsson P, Folke C, Crona B, Bodin O (2013) A theory of transformative agency in linked social-ecological systems. *Ecology and Society* **18**(3), 27. doi:10.5751/ES-05072-180327

Whyte KP, Brewer JP II, Johnson JT (2016) Weaving Indigenous science, protocols and sustainability science. *Sustainability Science* **11**, 25–32. doi:10.1007/s11625-015-0296-6

Wilson S (2008) *Research is Ceremony: Indigenous Research Methods*. Fernwood Publishing, Black Point, Canada.

Wyborn CA (2015) Connecting knowledge with action through coproductive capacities: adaptive governance and connectivity conservation. *Ecology and Society* **20**(1), 11. doi:10.5751/ES-06510-200111

4

Integrating development studies and social-ecological systems thinking: towards livelihood adaptation pathways

James R.A. Butler, Liana J. Williams, Toni Darbas, Tanya Jakimow, Kirsten Maclean and Clemens Grünbühel

Many of today's pressing human development challenges such as food insecurity, poverty, climate change and deforestation are embedded in systems characterised by multiple components, drivers of change emanating from different scales and sectors and with multiple equilibrium points, feedback loops and thresholds where abrupt and irreversible change can occur (Scoones 2009). As a consequence, Ramalingam (2013) has argued that the aid programs' established assumptions of linear, simple cause-and-effect relationships which have long guided interventions and their evaluation are no longer valid. In Chapter 7, we suggest that research designed to investigate complex development problems now needs to meet a 'hierarchy of needs', within which systems understanding is fundamental.

While various conceptual approaches to characterising complex systems in development are emerging (e.g. Reed *et al.* 2013; Dorward 2014), there are two mainstream paradigms which are applied widely. First is the Sustainable Livelihoods Framework (SLF), which integrates social science concepts from anthropology, economics, political ecology and human geography to understand the determinants of poverty and disadvantage, usually from the perspective of the household economy (Ellis and Biggs 2001; Kothari and Minogue 2002; Pieterse 2010). The SLF emerged in the 1990s from livelihoods analysis, which focuses on 'the capabilities, assets (including both material and social resources) and activities for a means of living' (Chambers and Conway 1992, p. 35). A core facet is the recognition of institutions, defined as 'systems of established and prevalent social rules that structure social interaction' (Hodgson 2006, p. 2), which, through power, politics and rights, mediate access to resources (Leach *et al.* 1999). Another is the concept of 'capitals' or 'assets' as inputs, which yield livelihood strategies and outcomes (Scoones 2009). The SLF has since become the participatory tool-of-choice among development practitioners in the analysis of disadvantage within poor communities (Carney 1998; Scoones 1998).

Second is social-ecological systems (SES), or resilience-thinking, which focuses on the linkages between natural resources and the ecosystem services that they provide to resource-dependent communities and related stakeholders (Walker *et al.* 2004; Walker and Salt 2006, Folke *et al.* 2010). An SES' state is determined by drivers and 'controlling variables' with thresholds which, if crossed, result in non-linear change to an alternate state. Identifying these variables and their thresholds is key to either avoiding them (hence maintaining system resilience) or intentionally crossing them to reach a more desirable state (transformation). Walker et al. (2010, p. S-17) have argued that this framework can usefully diagnose development challenges and identify interventions: 'in some circumstances, resilience may hinder escape from a poverty trap to a more desirable state [and] ... transformation has to be actively sought. Enhancing transformability is a major need in the developing world.' However, the inherent complexity of the SES framework, which requires significant amounts of data to enable analysis (Nelson et al. 2010a) and which uses ambiguous and abstract terminology, limits its applicability within participatory research (Leach 2008; McMurry 2010; Béné et al. 2011).

A growing body of research is focusing on how to operationalise systems analysis into stakeholder decision-making to achieve transformation in SESs (e.g. Abel et al. 2016; Butler et al. 2016a; O'Connell et al. 2016). With rapid and unprecedented global change, which is compounded by climate change (Scoones et al. 2007; Leach 2008; Ramalingam 2013), such decisions must also account for future uncertainty and unprecedented shocks to systems. 'Adaptation pathways' is the actor-based practice of sequencing decisions iteratively to plan development trajectories, while minimising the risks of maladaptation (Wise et al. 2014) and achieving climate-compatible development (Butler et al. 2016b). However, because adaptation pathways are intertwined with questions surrounding the transformation of societal progress towards sustainable development, the process is inherently political and requires the analysis of actors, their decision-making context and the power dynamics among them (Eisenhauer 2016). In addition, to achieve transformation, the underlying drivers of community or household vulnerability must be addressed; these are often the entrenched institutions and norms controlled by powerful vested interests (Lemos et al. 2007; Pelling 2011; Rodima-Taylor et al. 2012; Barrett and Constas 2014). In this sense, some of the complementary attributes of the SLF and SES frameworks could be combined to contribute to adaptation pathways: the SLF recognises the influence of power dynamics and institutional constraints on livelihood outcomes, while the SES identifies underlying variables and thresholds which control system states and impede transformation (Butler et al. 2014).

In spite of their parallel evolutions, there has been no attempt to assess the relative utility of the SLF and SES frameworks for understanding complex development problems, or designing appropriate interventions. This chapter has three objectives. First, we seek to explore the relative attributes of each framework by applying them to the same case study, a rural village in Thailand. Second, based on the results of the comparison, we present a hybrid framework which integrates their strengths to assess potential future development trajectories for the case study. Third, we discuss the contributions that this hybrid framework can make to constructing livelihood adaptation pathways, in terms of identifying key decision-making actors, power relations and dynamics among them, and underlying constraints to transformation.

Livelihoods analysis and the SLF

Livelihoods analysis and the SLF originated at the Institute of Development Studies in the early 1990s, where Chambers and Conway (1992) argued for an alternative to income-based measures of well-being, and put forward a framework that linked capability, equity and sustainability as central to understanding development. Development agencies began applying the SLF from the mid-1990s (Solesbury 2003; Toner 2003), and further refinement of the framework occurred in parallel (Scoones 1998; Carney 1998; Ellis 2000).

The framework analyses households' access to resources and capitals, as mediated by institutions and employed in the pursuit of different livelihood strategies. The resulting portfolio of activities is taken as determining livelihood outcomes for households and their members. These outcomes are to be understood within a particular political, social, economic and ecological context (Ellis 2000; Scoones 1998). Carney's SLF (Carney 1998) added the breakdown of resources into five capitals: financial, human, natural, social and physical, which may be substitutable. Others have argued that cultural and political capital should be added as distinct resources and inputs to livelihoods, to avoid framing these aspects purely as constraints (e.g. Baumann 2000; Daskon and McGregor 2012).

Scoones' (1998) framework poses the following question:

> ...given a particular **context** (of policy-setting, politics, history, agro-ecology and socio-economic conditions), what combination of **livelihood resources** (different types of 'capital') result in the ability to follow what combination of **livelihood strategies** (agricultural intensification/extensification, livelihood diversification and migration) with what **outcomes**? Of particular interest in this framework are the **institutional processes** (embedded in a matrix of formal and informal institutions and organisations) which mediate the ability to carry out such strategies and achieve (or not) such outcomes (p. 3, emphasis in original).

In this way, resource endowments of different groups or individuals are categorised and measured, with the ensuing portrait of strengths and weaknesses used to inform development interventions. Institutions are understood as able to represent bottlenecks or potential mechanisms to deliver 'sustainable livelihoods', defined as the resilience to recover from stresses and shocks and maintain or enhance capabilities and assets now and in the future, while not undermining the natural resource base (Ellis 2000).

Proponents outline several strengths of the SLF. First, it offers a contextually rich understanding of livelihoods, one that is holistic and locally embedded but which makes the connections between micro and macro influences (Rigg 2007; Scoones 2009). Second, it has a commitment to participatory methodologies that place people and their empowerment at the centre of the analysis (Chambers and Conway 1992; Bebbington 1999). Third, it highlights the role of institutions, power and politics in shaping individual and household livelihoods, historical inequality, exclusion from markets and the incompatibility of local institutions with dominant government policies (Scoones and Wolmer 2003). Finally, its adoption across organisations demonstrates its value as a credible boundary object which can bring disparate community members together and overcome other stakeholders' disciplinary and professional divides (Scoones 2009).

The limitations of SLF are also notable. Critics argue that it presents a static snapshot that cannot account for how livelihoods change over time (De Haan and Zoomers 2005; Allison and Horemans 2006; Jakimow 2013). While a focus of the framework, unless appropriately unpacked, institutional aspects become a 'black box' (Jakimow 2013), underrepresenting markets and policies (Dorward *et al.* 2003). However, this is often due to a technocratic application by development practitioners rather than the framework *per se* (Scoones and Wolmer 2003; Rigg 2007). The SLF also focuses on short-term, incremental coping strategies that can maintain 'sustainable' livelihoods but ignore influences from higher scales. It can therefore overlook radical, transformative change potentially required to pre-empt impending shifts in global drivers such as climate change. Finally, its roots in social sciences underplay ecological factors, such as ecosystem processes underpinning ecosystem services (Scoones 2009).

SES and resilience-thinking

An SES includes societal and ecological subsystems in mutual interaction (Gallopin 1991). SESs are characterised by three core features: irreducible uncertainty, the potential for non-linear dynamics due to cross-scale reinforcing feedback loops that amplify interactions within the system, and hence potential for the emergence of sudden and unexpected outcomes. Based largely on ecological evidence, it is possible for a system to shift between alternative stable states, having crossed thresholds. Identifying the drivers and controlling variables and their thresholds is key to either maintaining system resilience, or intentionally crossing them to navigate towards a more desirable state (transformation).

Adaptability (or adaptive capacity) is central to maintaining resilience, because it represents the social and ecological capacity to adjust to changes in drivers and variables, and thereby continue within the current system state. Responses may adapt to a particular set of issues (specified resilience) or to all kinds of shocks, including novel ones (general resilience). Transformability is the capacity to purposefully cross thresholds into new system states. SESs are connected across scales (panarchy), and may undergo adaptive cycles from phases of conservation to release, re-organisation to growth, and back to conservation (Gunderson and Holling 2002), but these patterns are not always evident (Armitage and Johnson 2006; Walker et al. 2006).

Human activities are typically viewed as sources of pressure on SESs, but the adaptability of human actors is also central to facilitating resilience and transformation. Key attributes are effective leadership, learning, innovation and knowledge integration, and the features of social networks, which together can be mobilised to generate collaborative governance (adaptive co-management). Evaluating stakeholders' relative influence and roles in natural resource management and their mental models of reality is, then, a key step in any resilience assessment (Resilience Alliance 2011).

Some of the terms used in resilience-thinking have alternative, and confusing, interpretations (Gunderson and Holling 2002; Gallopin 2006; Smit and Wandel 2006; Ensor 2011). Resilience in terms of an SES is defined in ecological terms as 'the capacity of a system to absorb disturbance and reorganise while undergoing change so as to still retain essentially the same function, structure, identity and feedbacks' (Walker et al. 2004, p. 4). In the social sciences it translates as a psychological term that refers to the ability of individuals or communities to cope with external stresses and disturbances (Adger 2000; Templeton and Adger 2004), which is the framing applied by the SLF (Ellis 2000). In resilience-thinking, vulnerability can be abstractly viewed as a system's proximity to a threshold. However, it is also used to measure communities' social, political, economic and environmental characteristics which render them more susceptible to hazards and shocks (Miller et al. 2010). Vulnerability manifests as poverty, which is characterised by limited assets such as savings, education, health, land, housing, food and political empowerment (Ensor and Berger 2009), and is therefore an economically and politically induced human condition which the SES approach does not depoliticise (Cannon and Müller-Mahn 2010).

Adaptability or adaptive capacity is an equally ambiguous term (Ensor 2011). In resilience-thinking, it is an attribute of the social and ecological capacity to absorb shocks and maintain current system function. However, from a development perspective, adaptive capacity has an emphasis on the potential for social actors within a system to respond to drivers of change, and to shape and create changes in that system (Chapin et al. 2006). The determinants of adaptive capacity in the SLF include both livelihood assets, including health, educational, financial and information resources, and the institutional and political contexts which determine how these are made available and mobilised (Smit and

Wandel 2006). Adaptive capacity can therefore mitigate vulnerability (poverty) and enhance the ability of actors in an SES to either maintain a system in its current state (absorbing capacity) or to proactively transform to a new state (purposeful adaptive capacity; Ensor 2011).

Although resilience-thinking includes the social in the form of stakeholders' adaptive capacity and mental models, adaptive co-management, controlling variables and drivers, the full complexity of human behaviour and values, power, politics and rights requires greater recognition and integration (Brown and Westaway 2011; Armitage *et al.* 2012; Stone-Jovicich 2015). This is particularly necessary when marginalised stakeholders are involved (Nadasdy 2007), raising the question of 'resilience for whom?' (Lebel *et al.* 2006; Leach 2008).

Applying the two approaches to a case study

Our case study is Sang Saeng, a village in the Isan region of north-eastern Thailand. Details are based on studies by Grünbühel (2004) and Grünbühel *et al.* (2003, 2007). A brief summary is provided here, and a full description is given in Appendix 1. It is an historical analysis of the village circa 2000, and the context has since changed.

The Isan region is ethnically Lao, and has been marginalised by the Thai government for decades. Distance from major economic centres, a dry climate, poor soils and lack of development investment renders the village poor by Thai standards. However, traditional rice and livestock-growing practices, reciprocal agricultural labour arrangements and hunter-gathering for subsistence remain strong, underpinned by Buddhism and customary institutions, norms and values which promote kinship networks, reciprocity, egalitarianism and cultural identity. These institutions also provide social safety nets for the poorest households and the elderly. Population pressure and the need for cash income drive villagers to work as migrant labourers in Bangkok and other parts of Thailand, but they return to the village for important agricultural activities. The village also engages with the national market economy by growing some non-glutinous rice as a cash crop (Fischer-Kowalski *et al.* 2011).

There is an elected village head, usually from an influential household, who is formally recognised as part of the local government administration. The position consults on decision-making with elected representatives and an informal consultative group of household heads. Informal political and moral authority is held by the village temple, particularly the head monk (abbot). Traditional institutions and values are evolving due to cash, modern values, consumer goods and building materials, new ideas and knowledge brought by the returning labourers.

There is social stratification according to asset holdings. Land-poor households are poorer, and the emerging cash economy tends to increase this economic differentiation. The current agricultural system has reached its productive capacity because all suitable land has been cleared of native forest and is fully utilised. Without free and reciprocal labour, the current agricultural system will become unviable. However, the community is still maintaining traditional practices, relying on draught provided by buffalo. The condition of ecosystem services provided by forests, soil, surface and groundwater is declining. Food security is not guaranteed during the dry season, and climate change projections suggest that droughts will become more frequent.

To provide a comparison of the two frameworks, we applied each to analyse the case study. For the SLF, we applied Scoones' (1998) version. For the SES framework we used the Resilience Alliance Workbook for Scientists (Resilience Alliance 2007), which involves three steps: 1) define the scale of interest, 2) current attributes of the system, including variables, thresholds and potential alternative states, and 3) implications for management interventions. For both exercises, we addressed the question: 'with respect to improving the food security and livelihoods of the Sang Saeng village community, what are the primary constraints and challenges requiring intervention?'

The full analytical processes are depicted in Figs 4.1 and 4.2. Here we summarise the concluding stages of each analysis. For the SLF this is 'livelihood outcomes', which identifies the different options and trade-offs in achieving sustainable livelihoods. For the SES approach this is Step 3, 'implications for management interventions'.

Findings using the SLF

The SLF identifies different outcomes for sustainable livelihoods under five criteria: increased numbers of working days, reduced poverty, improved well-being and capabilities, enhanced livelihood adaptation and resilience, and ensured sustainability of the natural resource base. In Sang Saeng, temporary migrant labour maintains employment and cash income, and hence contributes to poverty reduction in households engaging with the formal national economy (Fig. 4.1). Returning migrants also provide labour during peak periods of agricultural activity, enhancing capabilities for all households. The traditional agricultural system also perpetuates cultural and social capital, which enhances kinship ties, reciprocity and thus community resilience. Households are foremost subsistence farmers, producing sufficient rice but facing dry season food scarcity, which necessitates market engagement. The manual labour involved in rice production, coupled with poor nutrition and a lack of public health services, suggests pressures on physical health (i.e. human capital). Migrants also improve food security for those remaining, as the ecosystem is less strained. Migrants take rice, but everything else is shared by those remaining.

Although current livelihood outcomes may be sufficient, there are few available resources to expand production or accommodate population growth. However, migration reduces pressure on natural resources, and the importation of alternative building resources reduces harvest pressure on remaining forest timber. While migration is a key strategy to account for population growth, it has mixed implications for achieving sustainable livelihoods. Migration was occurring before modernisation and people used to migrate for fishing or salt extraction during the dry season, thus it is not a new phenomenon. Those returning bring new resources and assets into the kin network, which secures livelihoods for the households concerned. However, it also creates wealth disparity and new consumerist patterns, undermining the traditional institutions and egalitarian values which provide a social safety net for the poorest and elderly, increasing their vulnerability. Interventions should bolster traditional institutions and thus community resilience, while conserving natural resources and intensifying mixed subsistence and cash agriculture in tune with traditional cultural aspirations, therefore generating more sustainable livelihood options within the village.

Findings using the SES and resilience-thinking

The scale of interest was the Sang Saeng village community, including migrants who identified the village as their home. Scales above were the district, provincial, national and Asian political and economic levels, and households were the scale below. The second step identified the controlling variables for the village: the Thai political attitude towards the

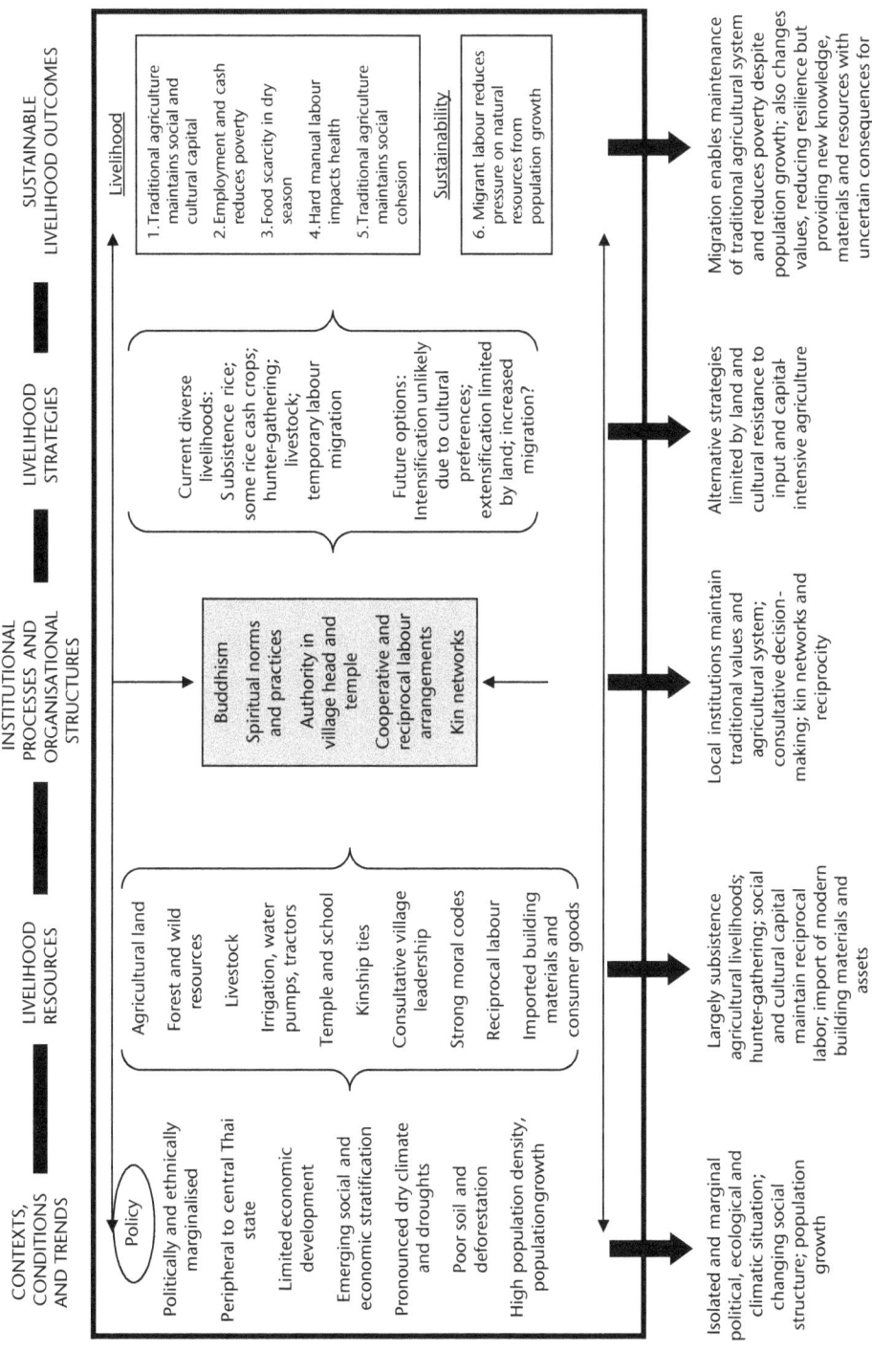

Fig. 4.1. Analysis of households' livelihoods within the Sang Saeng village applying the SLF.

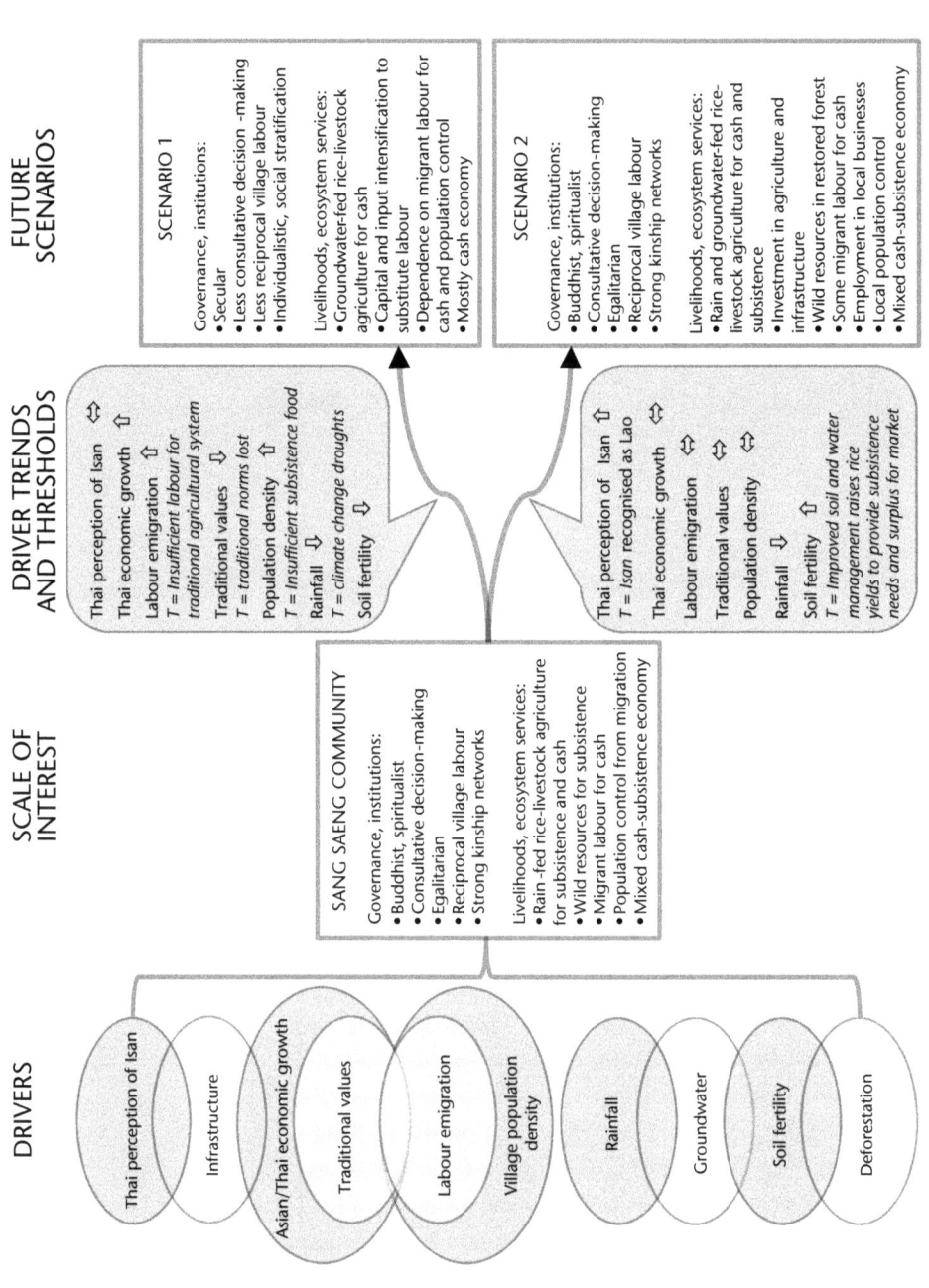

Fig. 4.2. Drivers of the Sang Saeng community (controlling variables shaded grey), its current characteristics, and two potential scenarios of system states due to trends (arrows) and thresholds (*T*) in drivers.

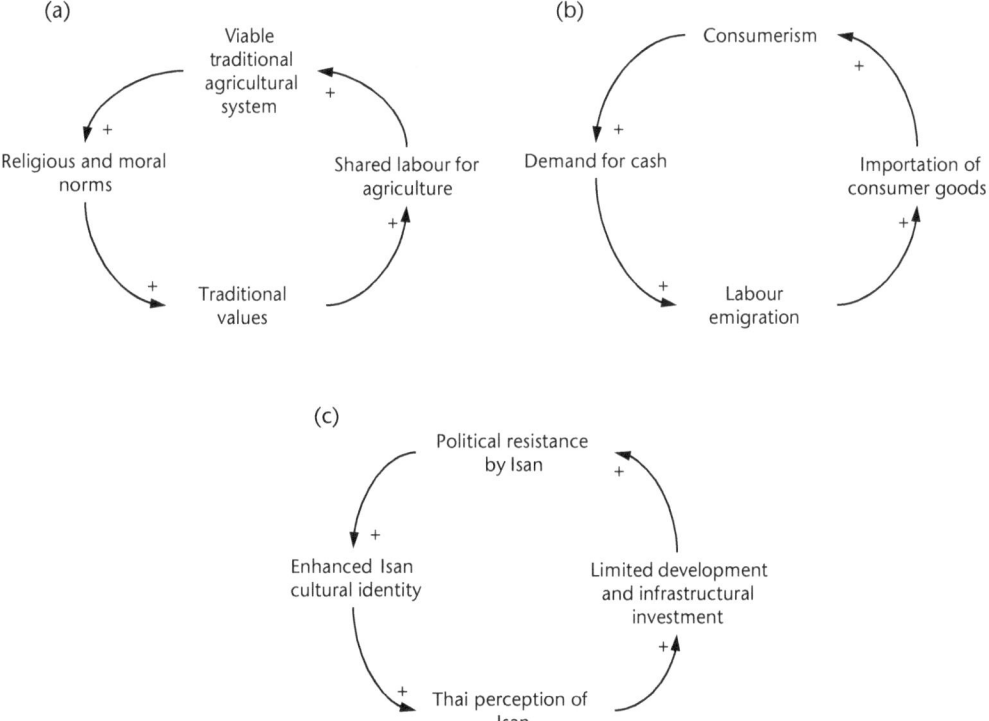

Fig. 4.3. Positive feedback loops among controlling variables (italicised) and other drivers of the Sang Saeng community: (a) traditional values, (b) labour emigration and (c) Thai perception of Isan. Feedback loop (a) is important for maintaining resilience of Sang Saeng today and in Scenario 2; (b) and (c) are relevant to maintaining Scenario 1.

Isan minority, Asian and Thai economic growth and linked labour markets, dilution of traditional values due to the influence of returning migrants, village population growth and density, which is also linked to migration, rainfall and drought, and soil fertility (Fig. 4.2).

The current Sang Saeng system is losing resilience, and may be approaching a 'release' phase in the adaptive cycle as the controlling variables of increasing labour emigration, declining traditional values, increasing population density and frequency of climate change-induced droughts approach cascading thresholds (Fig. 4.2). The traditional values which generate a positive feedback loop and maintain resilience today (Fig. 4.3a) are being changed by imported influences from returning labourers, which reduce adaptive capacity but introduce new knowledge, financial resources and innovation.

Consequently, transition to a new system state, Scenario 1, is most likely (Fig. 4.2). This development trajectory will result in a shift to a cash-based economy and capital- and input-intensive agriculture which is dependent on remittances provided by temporary and permanently migrated labour, as current traditional practices will become unviable without reciprocal labour. Traditional institutions will give way to more secular and individualistic behavioural patterns, reinforced by a positive feedback loop based on a demand for cash driven by labour emigration (Fig. 4.3b). The Thai perception of the Isan will remain negative, maintaining a positive feedback loop which exacerbates the political isolation of the community (Fig. 4.3c). This is occurring even as we write (in late 2016). This new state may not be stable, however, since community adaptive capacity will be weak, the condition of ecosystem services will continue to decline, and shocks from droughts may become more frequent.

A second and more desirable system state, Scenario 2, depends on a reversal of Thai perceptions of Isan. This results in increased development investment, which enables improved agricultural productivity and local business, countering the necessity for labour emigration. Consequently, traditional governance and kinship networks are maintained (Fig. 4.2), creating a reinforcing feedback loop (Fig. 4.3a). This state would be challenged by climate change-induced droughts, but would be more resilient due to the greater adaptive capacity provided by strong traditional institutions, innovation, healthy ecosystems and diverse livelihood activities.

The key challenge is how to avoid Scenario 1, by maintaining resilience of the current system. This requires some control of the influence of returning migrant labourers to limit the dilution of traditional institutions and values, while harnessing the benefits of cash, new ideas and innovation. Alternatively, proactive efforts could be made to transform to Scenario 2. This requires a shift in Thai attitudes towards the Isan, in order to provide development investment or to secure it from non-government organisations or international donors. Without development investment, it will be difficult to raise the current limits on agricultural production presented by labour, land and inputs. The balancing of traditional and modern values and knowledge within a mixed cash and subsistence economy will be necessary to optimise adaptive capacity, which will be important given impending climate change impacts. Stakeholders who will be central to enabling this transformation are the village head and temple monks, the consultative village decision-making forum, sympathetic provincial and national Thai government officials and politicians, plus non-government and international actors.

Comparison of the frameworks

Our comparison illustrates that both the SLF and SES frameworks are broadly similar in their fundamental aim of understanding and diagnosing a coupled human–environment system, and the options and trade-offs which inform interventions. However, there are some clear differences between their scope and areas of focus (Table 4.1). The SLF's priority is that 'people matter, contexts are important, a focus on capacities and capabilities rather than needs, and a normative emphasis on poverty and marginality' (Scoones 2009, p. 13). By comparison, the SES is concerned with system dynamics and potential future system states, and is normative towards the relative desirability of these outcomes for human well-being. The SLF focuses on households, and local formal and informal institutions that mediate capitals. The SES has an interest in multiple scales and interactions between them, the scale of interest being the community, and the governance of natural resources including stakeholders with key leadership roles.

These differences are exemplified by the concluding steps of the SLF and SES analyses. Both frameworks identify the same fundamental tension and paradox: the social 'costs' of returning migrant labourers on traditional institutions, values and the integrity of the largely subsistence agricultural and hunter-gatherer system, versus the benefits that they bring in terms of reciprocal agricultural labour, cash income, new materials and knowledge, and their mitigating effect on population pressure and food insecurity.

Nonetheless, there are differences in the interventions identified (Table 4.1). The SLF highlights the bolstering of traditional institutions which act as a safety net for poorer

Table 4.1. Comparative attributes of the SLF and the SES approaches, and resultant interventions identified for Sang Saeng

	SLF	SES
Focus	Poverty, vulnerable individuals and households and their livelihood strategies	Coupled human–environment systems, and actors required to manage them
Scale of interest	Individuals and households	Community or other scales (e.g. region, catchment)
Time horizon	Static: current and near-term	Dynamic: current and future
Principal concepts	Institutions, power dynamics, capitals, assets and capabilities	Drivers and controlling variables, feedback loops, thresholds, shocks, adaptability and transformability
Objectives	Resilient and sustainable livelihoods	Desired system state
Sang Saeng interventions	• Bolster traditional informal institutions as a social safety net for the poorest households and individuals • Introduce culturally appropriate agricultural intensification for subsistence and cash income	• Manage the influence of migrant labour on traditional institutions and values • Transform to Scenario 2 by addressing Thai alienation of Isan through actors at multiple scales

households and the elderly, while combining the conservation of natural resources with culturally appropriate intensification of agricultural production, thus generating more sustainable livelihood options. The SES focuses on managing key cross-scale variables (e.g. the Thai perception of Isan), thresholds and potential future system states.

Hence, the SES analysis identified two options. First, similar to the SLF's conclusion, the resilience of the current system could be enhanced by managing the impact of returning migrants on traditional institutions and values. Second, the system could be transformed to Scenario 2, which provides greater adaptive capacity for absorbing future perturbations such as climate change. This requires mobilisation and cooperation between key stakeholders to change Thai perceptions of the Isan at the national scale, resulting in increased development investment and a more productive agricultural system which integrates the benefits of traditional and modern values and resources. Thus the SLF identifies opportunities for creating sustainable livelihoods for households based on current capabilities, while the SES approach provides a future-oriented and anticipatory identification of critical multiscale interventions, and the actors required to implement them.

The SLF can enrich SES analysis

Our analysis of Sang Saeng is superficial and only for illustrative purposes. In spite of this, the comparison highlights two primary areas where the SLF could enhance the SES approach. First, the analysis of institutions and how they are structured and evolve is a key inquiry for the SLF (Jakimow 2013). Institutional analysis investigates the way in which informal (e.g. reciprocal labour arrangements) and formal institutions (e.g. government policies) regulate and constrain individuals' behaviour, and in turn shape governance of people and resources. Institutions occur at multiple levels, from the macro, which encompasses top-down structural processes (e.g. government policy), to the micro, which includes individuals' agency-based constructs (Leach *et al.* 1999). Exposing the characteristics, historical evolution and trajectories of institutions can also enable a practical-political economy, whereby development practitioners can proactively assist the poor to improve their conditions by capitalising on cultural change which modifies institutions (Li 1996). Such insights could contribute to a deeper understanding of institutional evolution as a driver in SESs, and related thresholds. Understanding the influence of power dynamics among stakeholders on the effectiveness of adaptive co-management, a potential oversight in resilience-thinking (Lebel *et al.* 2006; Nadasdy 2007; Cannon and Müller-Mahn 2010), would be an additional contribution.

Second, the SLF provides a useful framework for evaluating households' resources and capitals. For example, Nelson *et al.* (2010b) applied the concept to evaluate the relative adaptive capacity of agricultural communities in rural Australia. The Resilience Alliance Workbook for Scientists (Resilience Alliance 2007) also suggests the use of the capitals to evaluate adaptability. However, neither considered the critical relevance of institutions and institutional change as barriers and enablers in the mobilisation of capitals. Also, the role of cultural and political capital as distinct resources contributing to adaptive capacity should be considered. This is particularly relevant in developing country contexts where cultural identity is a key facet of resilience, and vulnerability is often perpetuated by a lack of political connections (Butler *et al.* 2014).

In turn, the SES provides three clear strengths. First, it explicitly exposes a broad array of social, economic, cultural and ecological controlling variables occurring at multiple scales, and therefore places livelihoods within a wider, more dynamic context. This is enhanced by identification of potential thresholds in cross-scale drivers and the types of shocks which may force thresholds to be breached, including ecosystem processes and services which are largely overlooked by the SLF. Second, the SES approach identifies potential long-term development scenarios and the feedback loops which may maintain these alternative system states. Third, taken together these facets can identify strategies required to manage key variables, thresholds and feedback loops in order to achieve system resilience or transformation. In addition, this process highlights the status of adaptive capacity and the key actors required to implement interventions.

Using Sang Saeng, it is possible to illustrate how the complementary concepts of the two frameworks could be integrated (Fig. 4.4). The SES approach's understanding of 'resilience' is applied (i.e. 'the capacity of a system to absorb disturbance and reorganise while undergoing change so as to still retain essentially the same function, structure, identity and feedbacks'; Walker *et al.* 2004, p. 4). 'Vulnerability' is used in the normative sense, as the social, political, economic and environmental context of communities which render them susceptible to hazards or shocks, mediated by their adaptive capacity (Miller *et al.* 2010). 'Adaptive capacity' is defined as the potential for actors within a system to respond to drivers, and to shape and create changes in that system (Chapin *et al.* 2006; Ensor 2011), including the reduction of vulnerability. We present our hybrid framework as step-wise questions.

Step 1: What are the characteristics of current livelihoods? From the SES, this step defines the system of interest, which may be the community or households within it, and the relevant scales above and below. The SLF, with its focus on poverty and disadvantage, contributes its second and third stages of analysis (capitals and institutions), and thus identifies the most vulnerable. The inclusion of political and cultural capital is an important addition, which can further capture power and institutional differentials within the community. The SES provides the concept of ecosystem services, which are either currently or potentially derived from natural capital identified from the SLF. Through this analysis it may be possible to identify key feedback loops from the SES approach which perpetuate disadvantage at the community or household scale.

Step 2: What are the drivers influencing livelihoods? From the SES approach, this step identifies multiscale controlling variables, including ecosystem processes. The SLF adds a detailed focus on institutional and cultural change. Information from the first stage of the SLF (context, conditions and trends) can also contribute to this step, by including institutional, policy and historical analyses.

Step 3: What are the trends in drivers and potential thresholds? Utilising the same process for SES, this investigates potential trends in controlling variables, thresholds and shocks which may cause multiple thresholds to be crossed, with a focus on cross-scale effects and thus the resilience attributes of the current system. General rather than specified resilience is the focus. The institutional, policy and historical analyses from Step 2 would contribute to forecasting potential trends and thresholds in social, economic and cultural drivers.

Step 4: What are the potential future scenarios for livelihoods? Based on alternative outcomes from shifts in drivers' trends and the crossing of thresholds, this step utilises the SES approach to envision alternative future system states, or scenarios. The understanding of feedback loops from Step 1 will assist in understanding how resilient the alternative systems may be. From the SLF, the alternative scenarios can be normatively defined in terms of the reduction of poverty and vulnerability and enhanced livelihood sustainability, hence a desired development trajectory can be identified as an objective. From the SES approach, the relative importance of maintaining current system resilience or transformation can then be considered.

Step 5: What is the adaptive capacity of the community and/or households today? This step analyses the current adaptive capacity of the community and/or households plus linked actors to respond to current drivers, tackle vulnerability and steer the system towards the desired state (for Sang Saeng, Scenario 2). This combines the SLF concepts of capitals and mediating institutions analysed in Step 1 with the SES perspectives on adaptability, such as the capacity for learning, leadership, innovation and social networks. Some of these attributes (e.g. leadership) overlap with the concept of capitals (e.g. social capital) from the SLF. Ecological perspectives on adaptability can also be considered, such as habitat connectivity and remnant vegetation, which are a component of natural capital.

Step 6: What livelihood adaptation strategies are necessary? This step concludes the process by identifying multiscale interventions in the form of adaptation strategies required to proactively transform the community and/or vulnerable households towards Scenario 2. From the SES approach, this requires the identification of influential actors who can collaborate to initiate adaptive co-management and mobilise the attributes of adaptive capacity determined in Step 5. The SLF provides the identification of livelihood strategies (intensification or extensification, diversification or migration), plus an understanding and pre-empting of the power and politics which will influence decision-making, potentially further marginalising disadvantaged households or individuals. In addition, appropriate institutional analysis from earlier steps could enable a practical-political economy (after Li 1996), whereby the poor are assisted to capitalise on cultural change to reduce their vulnerability.

This hybrid framework can also provide the basis for constructing livelihood adaptation pathways. For the first key adaptation pathways component – the identification of actors and their decision-making context – the SLF emphasises poorer individuals' or households' capacities and capabilities, and the decisions they are able to take to achieve livelihood outcomes (Table 4.2). The SES approach gives more emphasis to stakeholders at different and higher scales of the system who can implement interventions to tackle

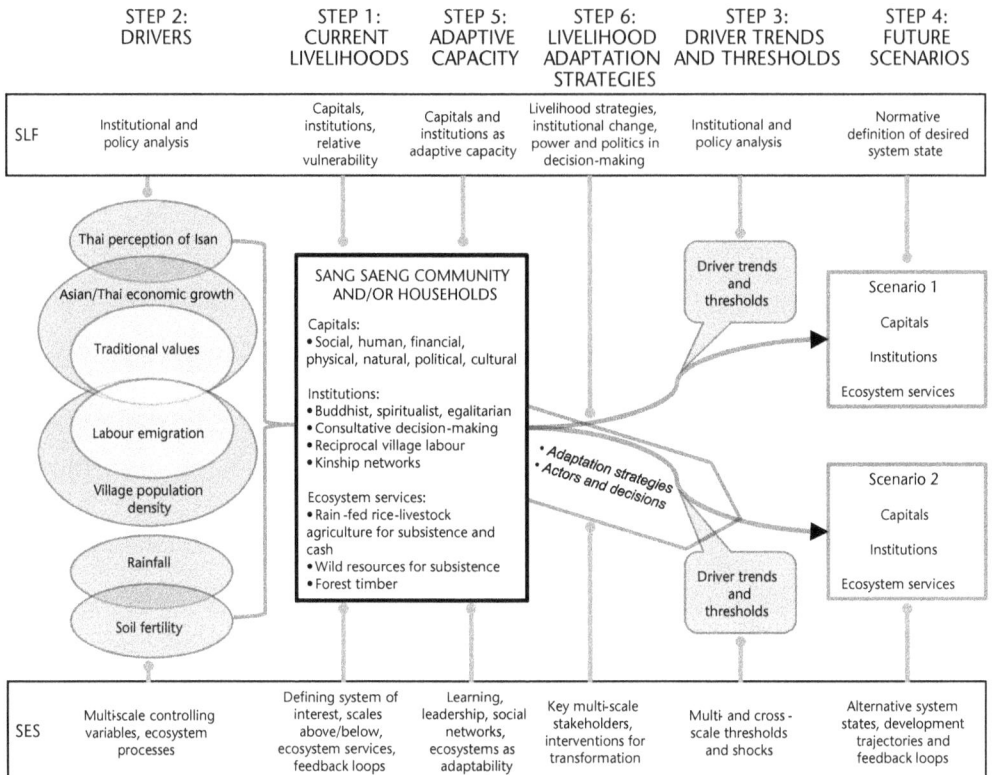

Fig. 4.4. Proposed integration of complementary concepts from the SLF and SES approach into livelihood adaptation pathways, applying aspects of the Sang Saeng case study.

controlling variables. For the second component – the recognition of power relations and dynamics among decision-makers – the SLF has a strong focus on institutions and political capital, and their influence on marginalising or favouring individuals or households. The SES approach may only consider power and politics as a controlling variable, and does not deconstruct the characteristics or evolution of institutions as the SLF could. For the third component – understanding the underlying barriers to transformation – the SLF again is strong due to its emphasis on formal and informal institutions. However, the SES framework adds value by considering these issues as dynamic variables with thresholds that can be managed. In combination, these processes could identify transformative livelihood adaptation strategies in Step 6 of our hybrid framework, plus the actors and decisions required to implement them (Fig. 4.4).

Conclusion

The application of our hybrid framework within participatory decision-making processes may meet mixed success (Butler *et al.* 2016c). The SLF's key concepts are proven to be credible and accessible to community members and multiple stakeholders (Scoones 2009). By contrast, experience of using the SES approach as a participatory tool suggests that resilience-thinking concepts are not easily translated in community settings where limited capacity and language barriers may inhibit the explanation required (Béné *et al.* 2011). Participants may also have culturally determined perceptions of the future which are not easily reconciled with conceptualisations of future scenarios (Wollenberg *et al.* 2000;

Table 4.2. Contrasting features of the SLF and the SES approaches in terms of three components central to adaptation pathways

Component	SLF	SES
1. Actors and their decision-making context	Focus on poorer individuals' and households' capacities and capabilities (e.g. *sustainable livelihood outcome options for resource-dependent households*)	Identification of stakeholders across scales who can implement interventions and tackle controlling variables (e.g. *sympathetic provincial and national government officials and non-government and international actors who can shift Thai attitudes towards the Isan to provide development investment*)
2. Power relations and dynamics	Focus on political capital and institutions which influence social and economic stratification at the individual and household scale (e.g. *power of the village head and political capital of related households; informal consultative decision-making forum*)	Identification of power and politics as a controlling variable at different scales; explores potential changes and thresholds in power as a driver, and key stakeholders who influence adaptability (e.g. *Thai political influence on Isan and development; collaboration between village head, temple monks and government actors to implement strategies*)
3. Underlying issues constraining transformations	Focus on formal and informal institutions which mediate livelihood outcomes (e.g. *influence of migrant labour on traditional institutions and values; cultural and financial capital limitations to intensifying agricultural productivity*)	Identification of cross-scale controlling variables (e.g. *Thai perceptions of Isan; population pressure; migrant labour changing traditional institutions and values*)

Examples from Sang Saeng are given in italics.

Bohensky et al. 2016). Challenges with abstract and ambiguous terminology are also likely to occur when engaging policy stakeholders (Leach 2008). Further, placing researchers and their scientific concepts in dominant positions in such situations can influence equitable knowledge integration and hence the outcomes of participatory processes (Folke and Fabricius 2005; Fazey et al. 2010; Gratani et al. 2011).

Therefore, we suggest that the true utility of our hybrid SLF–SES framework and its terminology and concepts can be fully determined only once tested and perhaps simplified within participatory processes. As it stands, it may remain an exposition of the potential synergies between development studies and resilience-thinking, rather than a tool which can be applied in the practice of livelihood adaptation pathways. As such it will perpetuate the existing flaw in the SES approach – its complexity and abstract terminology render it useful only to scientists and practitioners diagnosing development challenges, in isolation from the subjects it intends to benefit. Hence, as well as its contributions to an enhanced understanding of power dynamics and institutions, the SLF and development studies can inform the design and implementation of decision-making processes necessary for livelihood transformation.

Acknowledgements

Luis Rodriguez and Lucy Carter provided helpful reviews of earlier drafts. Jack Katzfey provided down-scaled climate projections for the region.

References

Abel N, Wise RM, Colloff M, Walker B, Butler JRA, Ryan P, Norman C, Langston A, Anderies J, Gorddard R, Dunlop M, O'Connell D (2016) Building resilient pathways to transformation when no-one is in charge: insights from Australia's Murray-Darling Basin. *Ecology and Society* 21, 23. doi:10.5751/ES-08422-210223

Adger WN (2000) Social and ecological resilience: are they related? *Progress in Human Geography* 24, 347–364. doi:10.1191/030913200701540465

Allison EH, Horemans B (2006) Putting the principles of the Sustainable Livelihoods Approach into fisheries development and policy. *Marine Policy* 30, 757–766. doi:10.1016/j.marpol.2006.02.001

Armitage D, Johnson D (2006) Can resilience be recognised with globalization and increasingly complex resource degradation in Asian coastal regions? *Ecology and Society* 11, 2.

Armitage D, Béné C, Charles AT, Johnson D, Allison EH (2012) The interplay of well-being and resilience in applying a social-ecological perspective. *Ecology and Society* 17, 15. doi:10.5751/ES-04940-170415

Barrett C, Constas M (2014) Toward a theory of resilience for international development applications. *Proceedings of the National Academy of Sciences of the United States of America* 111, 14625–14630. doi:10.1073/pnas.1320880111

Baumann P (2000) *Sustainable Livelihoods and Political Capital: Arguments and Evidence from Decentralisation and Natural Resource Management in India*. Working Paper. Overseas Development Institute, London.

Bebbington A (1999) Capitals and capabilities: a framework for analyzing peasant viability, rural livelihoods and poverty. *World Development* 27, 2021–2044. doi:10.1016/S0305-750X(99)00104-7

Béné C, Evans L, Mills D, Ovie S, Raji A, Tafida A, Kodio A, Sinaba F, Morand P, Lemoalle J, Andrew N (2011) Testing resilience thinking in a poverty context: experience from the Niger River basin. *Global Environmental Change* 21, 1173–1184. doi:10.1016/j.gloenvcha.2011.07.002

Bohensky EL, Kirono D, Butler JRA, Rochester W, Habibi P, Handayani T, Yanuartati Y (2016) Climate knowledge cultures: stakeholder perspectives on change and adaptation in Nusa Tenggara Barat, Indonesia. *Climate Risk Management* 12, 17–31. doi:10.1016/j.crm.2015.11.004

Brown K, Westaway E (2011) Agency, capacity and resilience to environmental change: lessons from human development, well-being, and disasters. *Annual Review of Environment and Resources* 36, 321–342. doi:10.1146/annurev-environ-052610-092905

Butler JRA, Suadnya W, Puspadi K, Sutaryono Y, Wise RM, Skewes TD, Kirono D, Bohensky EL, Handayani T, Habibi P, Kisman M, Suharto I, Hanartani, Supartarningsih S, Ripaldi A, Fachry A, Yanuartati Y, Abbas G, Duggan K, Ash A (2014) Framing the application of adaptation pathways for rural livelihoods and global change in Eastern Indonesian islands. *Global Environmental Change* 28, 368–382. doi:10.1016/j.gloenvcha.2013.12.004

Butler JRA, Bohensky EL, Darbas T, Kirono DGC, Wise RM, Sutaryono Y (2016a) Building capacity for adaptation pathways in eastern Indonesian islands: synthesis and lessons learned. *Climate Risk Management* 12, A1–A10. doi:10.1016/j.crm.2016.05.002

Butler JRA, Bohensky EL, Suadnya W, Yanuartati Y, Handayani T, Habibi P, Puspadi K, Skewes TD, Wise RM, Suharto I, Park SE, Sutaryono Y (2016b) Scenario planning to leap-frog the Sustainable Development Goals: an adaptation pathways approach. *Climate Risk Management* 12, 83–99. doi:10.1016/j.crm.2015.11.003

Butler JRA, Suadnya IW, Yanuartati Y, Meharg S, Wise RM, Sutaryono Y, Duggan K (2016c) Priming adaptation pathways through adaptive co-management: design and evaluation for developing countries. *Climate Risk Management* **12**, 1–16. doi:10.1016/j.crm.2016.01.001

Cannon T, Müller-Mahn D (2010) Vulnerability, resilience and development discourses in context of climate change. *Natural Hazards* **55**, 621–635. doi:10.1007/s11069-010-9499-4

Carney D (1998) *Sustainable Livelihoods Approaches: Progress and Possibilities for Change.* Overseas Development Institute, London.

Chambers R, Conway G (1992) *Sustainable Rural Livelihoods: Practical Concepts for the 21st Century.* IDS Discussion Paper. Institute of Development Studies, Brighton, UK.

Chapin FS, Lovecraft AL, Zavaleta ES, Nelson J, Robards MD, Kofinas GP, Trainor SF, Peterson GD, Huntingdon HP, Naylor RL (2006) Policy strategies to address sustainability of Alaskan boreal forests in response to a directionally changing climate. *Proceedings of the National Academy of Sciences of the United States of America* **103**, 16637–16643. doi:10.1073/pnas.0606955103

Daskon C, McGregor A (2012) Cultural capital and sustainable livelihoods in Sri Lanka's rural villages: towards culturally aware development. *Journal of Development Studies* **48**, 549–563. doi:10.1080/00220388.2011.604413

De Haan L, Zoomers A (2005) Exploring the frontiers of livelihoods research. *Development and Change* **36**, 27–47. doi:10.1111/j.0012-155X.2005.00401.x

Dorward AR (2014) Livelisystems: a conceptual framework integrating social, ecosystem, development, and evolutionary theory. *Ecology and Society* **19**, 44. doi:10.5751/ES-06494-190244

Dorward AR, Poole ND, Morrison JA, Kydd JG, Urey I (2003) Markets, institutions and technology: missing links in livelihoods analysis. *Development Policy Review* **21**, 319–332. doi:10.1111/1467-7679.00213

Eisenhauer DC (2016) Pathways to climate change adaptation: making climate change action political. *Geography Compass* **10**(5), 207–221. doi:10.1111/gec3.12263

Ellis F (2000) *Rural Livelihoods and Diversity in Developing Countries.* Oxford University Press, Oxford.

Ellis F, Biggs S (2001) Evolving themes in rural development 1950s–2000s. *Development Policy Review* **19**, 437–448. doi:10.1111/1467-7679.00143

Ensor J (2011) *Uncertain Futures: Adapting Development to a Changing Climate.* Practical Action Publishing, Rugby, UK.

Ensor J, Berger R (2009) *Understanding Climate Change Adaptation: Lessons from Community-based Approaches.* Practical Action Publishing, Rugby, UK.

Fazey I, Kesby M, Evely A, Latham I, Wagatora D, Hagasua J-E, Reed MS, Christie M (2010) A three-tiered approach to participatory vulnerability assessment in the Solomon Islands. *Global Environmental Change* **20**, 713–728. doi:10.1016/j.gloenvcha.2010.04.011

Fischer-Kowalski M, Singh SJ, Lauk C, Remesch A, Ringhofer L, Grünbühel C (2011) Sociometabolic transitions in subsistence communities: Boserup revisited in four comparative case studies. *Human Ecology Review* **18**, 147–158.

Folke C, Fabricius C (2005) Communities, ecosystems and livelihoods. In *Millennium Ecosystem Assessment: Ecosystems and Human Well-being Sub-global Assessments* pp. 261–276. Island Press, Washington DC.

Folke C, Carpenter SR, Walker B, Scheffer M, Chapin T, Rockström J (2010) Resilience thinking: integrating resilience, adaptability and transformability. *Ecology and Society* **15**, 20. http://www.ecologyandsociety.org/vol15/iss4/art20.

Gallopin GC (1991) Human dimensions of global change: linking the global and the local processes. *International Social Science Journal* **130**, 707–718.

Gallopin GC (2006) Linkages between vulnerability, resilience and adaptive capacity. *Global Environmental Change* **16**, 293–303. doi:10.1016/j.gloenvcha.2006.02.004

Gratani M, Butler JRA, Royee F, Burrows D, Valentine P, Canendo W, Anderson A (2011) Is validation of Indigenous ecological knowledge a disrespectful process? A case study of traditional fishing poisons and invasive fish management from the Wet Tropics, Australia. *Ecology and Society* **16**, 25. doi:10.5751/ES-04249-160325

Grünbühel CM (2004) Resource use systems and rural smallholders: an analysis of two Lao communities. PhD thesis. Faculty for Humanities and Social Sciences, University of Vienna, Austria.

Grünbühel CM, Haberl H, Schandl H, Winiwarter V (2003) Socio-economic metabolism and colonization of natural processes in Sang Saeng village: material and energy flows, land use, and cultural change in northeast Thailand. *Human Ecology* **31**, 53–86. doi:10.1023/A:1022882107419

Grünbühel CM, Singh S, Fischer-Kowalski M (2007) The local base of transitions in developing countries. In *Socioecological Transitions and Global Change: Trajectories of Social Metabolism and Land Use*. (Eds M Fischer-Kowalski, H Haberl) pp. 139–178. Edward Elgar, Cheltenham, UK.

Gunderson LH, Holling CS (Eds) (2002) *Panarchy: Understanding Transformations in Human and Natural Systems*. Island Press, Washington DC.

Hodgson GM (2006) What are institutions? *Journal of Economic Issues* **40**(1), 1–25. doi:10.1080/00213624.2006.11506879

Jakimow T (2013) Unlocking the black box of institutions in livelihoods analysis: case study from Andhra Pradesh, India. *Oxford Development Studies* **41**, 493–516. doi:10.1080/13600818.2013.847078.

Kothari U, Minogue M (Eds) (2002) *Development Theory and Practice: Critical Perspectives*. Palgrave, London.

Leach M (Ed.) (2008) Re-framing resilience: a symposium report. STEPS Working Paper 13. Social, Technological and Environmental Pathways to Sustainability (STEPS) Centre, Brighton, UK. <http://steps-centre.org/wp-content/uploads/Resilience.pdf>.

Leach M, Mearns R, Scoones I (1999) Environmental entitlements: dynamics and institutions in community-based natural resource management. *World Development* **27**, 225–247. doi:10.1016/S0305-750X(98)00141-7

Lebel L, Anderies JM, Campbell B, Folke C, Hatfield-Dodds S, Hughes TP, Wilson J (2006) Governance and the capacity to manage resilience in regional social-ecological systems. *Ecology and Society* **11**, 19. http://www.ecologyandsociety.org/vol11/iss1/art19/

Lemos MC, Boyd E, Tompkins EL, Osbahr H, Liverman D (2007) Developing adaptation and adapting development. *Ecology and Society* **12**, 26. http://www.ecologyandsociety.org/vol12/iss2/art26/

Li TM (1996) Images of community: discourse and strategy in property relations. *Development and Change* **27**, 501–527. doi:10.1111/j.1467-7660.1996.tb00601.x

McMurry A (2010) The rhetoric of resilience: in a world that runs on persuasion, the green movement would be wise to re-examine its use of language. *Alternatives Journal* **36**(2). <http://www.alternativesjournal.ca/magazines/362-buildingresilience>.

Miller F, Osbahr H, Boyd E, Thomalla F, Bharwani S, Ziervogel G, Walker B, Birkmann J, van der Leeuw S, Rockström J, Hinkel J, Downing T, Folke C, Nelson D (2010) Resilience and vulnerability: complementary or conflicting concepts? *Ecology and Society* **15**, 11. http://www.ecologyandsociety.org/vol15/iss3/art11/

Nadasdy P (2007) Adaptive co-management and the gospel of resilience. In *Adaptive Co-management: Collaboration, Learning, and Multi-level Governance*. (Eds D Armitage, F Berkes, N Doubleday) pp. 208–227. UBC Press, Vancouver.

Nelson R, Kokic P, Crimp S, Meinke H, Howden SM (2010a) The vulnerability of Australian rural communities to climate variability and change. Part I. Conceptualising and measuring vulnerability. *Environmental Science and Policy* **13**, 8–17. doi:10.1016/j.envsci.2009.09.006

Nelson R, Kokic P, Crimp S, Martin P, Meinke H, Howden SM, de Voil P, Nidumolu U (2010b) The vulnerability of Australian rural communities to climate variability and change. Part II. Integrating impacts with adaptive capacity. *Environmental Science and Policy* **13**, 18–27. doi:10.1016/j.envsci.2009.09.007

O'Connell D, Abel N, Grigg N, Maru Y, Butler JRA, Cowie A, Stone-Jovicich S, Walker B, Wise R, Ruhweza A, Pearson L, Ryan P, Stafford Smith M (2016) *Designing Projects in a Rapidly Changing World: Guidelines for Embedding Resilience, Adaptation and Transformation into Sustainable Development Projects*. Version 1.0. Global Environment Facility, Washington DC.

Pelling M (2011) *Adaptation to Climate Change: From Resilience to Transformation*. Routledge, London.

Pieterse N (2010) *Development Theory: Deconstructions/Reconstructions*. 2nd edn. Sage, London.

Qadir M, Noble AD, Chartres C (2013) Adapting to climate change by improving water productivity of soils in dry areas. *Land Degradation and Development* **24**, 12–21. doi:10.1002/ldr.1091

Ramalingam B (2013) *Aid on the Edge of Chaos: Rethinking International Cooperation in a Complex World*. Oxford University Press, London.

Reed MS, Podestá G, Fazey I, Geeson N, Hessel R, Hubacek K, Letson D, Nainggolan D, Prell C, Rickenbach MG, Ritsema C, Schwilch G, Stringer LC, Thomas AD (2013) Combining analytical frameworks to assess livelihood vulnerability to climate change and analyse adaptation options. *Ecological Economics* **94**, 66–77. doi:10.1016/j.ecolecon.2013.07.007

Resilience Alliance (2007) *Assessing Resilience in Social-ecological Systems: A Workbook for Scientists*. Version 1.1. <http://www.resalliance.org/index.php/resilience_assessment>.

Resilience Alliance (2011) *Applying a Social-ecological Inventory: A Workbook for Finding the Key Actors and Engaging Them*. <http://www.resalliance.org/index.php/resilience_assessment>.

Rigg J (2007) *An Everyday Geography of the Global South*. Routledge, London.

Rodima-Taylor D, Olwig MF, Chhetri N (2012) Adaptation as innovation, innovation as adaptation: an institutional approach to climate change. *Applied Geography* **33**, 107–111. doi:10.1016/j.apgeog.2011.10.011

Scoones I (1998) *Sustainable Rural Livelihoods: A Framework for Analysis*. Institute for Development Studies, Brighton, UK.

Scoones I (2009) Livelihoods perspectives and rural development. *Journal of Peasant Studies* **36**, 171–196. doi:10.1080/03066150902820503

Scoones I, Wolmer W (2003) Introduction: livelihoods in crisis. Challenges for rural development in southern Africa. *IDS Bulletin* **34**, 1–14. doi:10.1111/j.1759-5436.2003.tb00073.x

Scoones I, Leach M, Smith A, Stagl S, Stirling A, Thompson J (2007) *Dynamic Systems and the Challenge of Sustainability*. Working Paper 1. Social, Technological and Environmental Pathways to Sustainability (STEPS) Centre, Brighton, UK. <http://opendocs.ids.ac.uk/opendocs/handle/123456789/2470>.

Smit B, Wandel J (2006) Adaptation, adaptive capacity and vulnerability. *Global Environmental Change* **16**, 282–292. doi:10.1016/j.gloenvcha.2006.03.008

Solesbury W (2003) *Sustainable Livelihoods: A Case Study of the Evolution of DFID Policy*. Working Paper 217. Overseas Development Institute, London.

Stone-Jovicich S (2015) Probing the interfaces between the social sciences and social-ecological resilience: insights from integrative and hybrid perspectives in the social sciences. *Ecology and Society* **20**, 25. doi:10.5751/ES-07347-200225

Stuart-Fox M (1997) *A History of Laos*. Cambridge University Press, Cambridge.

Templeton EL, Adger WN (2004) Does adaptive management of natural resources enhance resilience to climate change? *Ecology and Society* **9**, 2.

Toner A (2003) Exploring sustainable livelihoods approaches in relation to two interventions in Tanzania. *Journal of International Development* **15**, 771–781. doi:10.1002/jid.1030

Walker BH, Salt D (2006) *Resilience Thinking: Sustaining Ecosystems and People in a Changing World*. Island Press, Washington DC.

Walker BH, Holling CS, Carpenter SC, Kinzig AP (2004) Resilience, adaptability and transformability. *Ecology and Society* **9**, 5.

Walker BH, Anderies JM, Kinzig AP, Ryan P (2006) *Exploring Resilience in Social-Ecological Systems: Comparative Studies and Theory Development*. CSIRO Publishing, Melbourne.

Walker B, Sayer J, Andrew NL, Campbell BM (2010) Should enhanced resilience be an objective of natural resource management research for developing countries? *Crop Science* **50**, S-10–S-19. doi:10.2135/cropsci2009.10.0565

Wise RM, Fazey I, Stafford Smith M, Park SE, Eakin HC, Archer van Garderen ERM, Campbell B (2014) Reconceptualising adaptation to climate change as part of pathways of change and response. *Global Environmental Change* **28**, 325–336. doi:10.1016/j.gloenvcha.2013.12.002

Wollenberg E, Edmunds D, Buck L (2000) Using scenarios to make decisions about the future: anticipatory learning for the adaptive co-management of community forests. *Landscape and Urban Planning* **47**, 65–77. doi:10.1016/S0169-2046(99)00071-7

Appendix 1

Case study: Sang Saeng village, north-east Thailand

Though considered by the Thai government as ethnically Thai, the majority of people in Isan are closer in linguistic and cultural tradition to the Lao of Lao PDR. This stems from a complex history of conflict and struggle for territory during the 16th to 19th centuries. The Lao Kingdom of Lan Xang, which spanned present-day Lao PDR, Isan and north-western Vietnam, disintegrated into three principalities in the late 1600s (Stuart-Fox 1997). The loss of unity and power resulted in the principalities falling under the hegemony of Siam (Thailand). The former territories of Lan Xang were ruled variously by the Siamese, present-day Vietnam and China until the mid-19th century. The Siamese largely exploited and neglected Isan settlements, and a strong sense of Lao rebellion and cultural identity persisted. The territorial friction culminated with the ceding of Isan to Thailand by the colonial power, France, in the early 1900s, leaving a larger population of ethnic Lao living outside the Lao territory than within it (Stuart-Fox 1997).

In the late 19th century, the Thai introduced a centralised administrative structure, removing power from local lords. Reforms were opposed in Isan and the region was largely neglected by the Thai government's development policies. Marginalisation, coupled with efforts to suppress Lao identity, fostered discontent. Leftist politics became active in Isan in the 1950s. Fears spread of a communist uprising in the region and in the 1960s the US government established military bases there to launch air raids on Laos and Vietnam; it also installed roads, dams and telecommunications. Strongly distinct cultural identities persisted, however, reinforced through Thai discrimination against migrant labourers from Isan.

This tension and the political dominance of the ethnic Thai has fostered strong political opposition within the Isan region, demonstrated by protests against the Pak Mun Dam in the 1990s and the March on Bangkok in 1992, which led to the 'Assembly of the Poor' in 1995. The region remains one of the poorest in Thailand and sits on the political and economic periphery. There are few major economic centres, and there has been little integration or engagement with development pursued through the national economy, fuelled by rapid growth in the Asian economy. However, through a national trend in decentralisation the Ubon Ratchathani provincial government has become politically recognised, augmented by non-government and international donors who highlight the development deficit in the Isan region.

The region is typified by poor-quality, sandy soils and a low average annual rainfall of 1183 mm (Qadir *et al.* 2013). Originally covered with semi-deciduous forest, deforestation to create agricultural land is widespread. Climatically, the region has a pronounced dry season from mid-December to mid-May and is prone to drought. Down-scaled climate projections under 'business as usual' global greenhouse gas emission scenarios suggest that droughts will become more frequent (J Katzfey, CSIRO, pers. comm.).

As of 2000, the population of Sang Saeng village is ~171 people within 50 households. Population density is 93 people/km^2, and all land suitable for agriculture is utilised. Although annual population growth has fallen to 1.37%, at this rate numbers could double by 2060. Of the 184 ha village area, 18% retains forest cover, largely because it is too far from water sources to be irrigated. The Mun River, the nearest major continuous water source, lies 20 km from the village. A public groundwater pump and storage facility was constructed by the government in 1994. Prior to that, water was collected manually and stored in jars. Irrigation channels are constructed and maintained by the community. A semi-permanent road built by the government links the village to the provincial road network. Other government services include electricity, a primary school and a health centre, which are shared with other villages in the sub-district.

The village is ethnically homogenous (*Thai-Isan*), and there is limited social stratification. However, there is wealth and land ownership stratification: of the 50 households, 14 are landless and another 14 own less than the average of 2.9 ha. Only two houses own hand tractors. In spite of this, social mobility is not limited by ethnicity or recent migration, as is often the case in other villages. Informal political and moral authority is held by the village temple (*wat*) and its monks, the village head and others with specialised knowledge used for the benefit of the community (e.g. teachers, spirit mediums or those who have spent time outside the village). Buddhism, belief in the spirit world (*phi*) and life force (*khuan*) provide a moral code for the villagers. Ceremonies and rituals mark transitions, placate spirits, engender a sense of solidarity and reaffirm the shared interests of the community. Festivals are also embedded within the agricultural calendar to mark the beginning of the planting season and ensure good harvests. The village head is elected, usually from an influential and well-off household. The position is formally recognised as part of the state administration. In addition, there is an informal group, comprising all the heads of household compounds, which the village head will consult before making decisions.

Households are often multigenerational; if one generation moves out, they will usually stay in close proximity and form family compounds. Households share agricultural labour to produce a common harvest. Notions of kinship extend beyond family, and reciprocal bonds among the community are strong. Land is privately owned and passed down to the youngest daughter, who is expected to look after her parents in their old age. While land ownership denotes exclusive use, any wild resources are communally owned. The felling of live trees has been banned by the government due to deforestation. Consequently, wood for construction and other purposes has become more expensive, triggering the import of modern materials such as concrete for building housing.

The village economy is dependent on rice farming, hunter-gathering of local resources and labour migration. Up to 30% of land is used to cultivate the state-promoted non-glutinous export rice, which is the main commercial production activity. Most glutinous rice, vegetables, chicken and duck is kept for home consumption. Rice yields are typically low, averaging 1.4 t/ha. With all viable agricultural land in use, rice paddy land is one of the most important resources in the village. With population growth, land parcels are reducing in size as families continue to subdivide through inheritance. Fields are permanent and largely rain-fed, with supplementary irrigation pumped from streams and canals. The two households that own hand tractors use buffalo as the preferred method of draught, due to their low maintenance costs and value as meat, capital and ceremonial use. Rice threshing and pumping of water are the main activities that have been mechanised. Buffalo are bred as working animals, but some households also raise cattle for market. Livestock are left to graze in fields during the dry season, providing organic fertiliser. Overnight and during cropping they are kept in stables where they are fed straw and grass. Manure is mixed with straw and applied to the fields to maintain otherwise-declining soil fertility. Households limit chemical fertiliser use due to its cost and the negative impacts on soil acidity, fish, insects and molluscs in the flooded rice fields, which they utilise for consumption.

Peak working periods are rice nursery preparation, ploughing, transplanting and harvest. During these times households work together and villagers who have migrated for work return to contribute their labour. At harvest, larger communal working groups are formed. During the dry season, hunting and foraging of fish, birds, amphibians, insects, leaves, flowers, herbs, mushrooms and roots are key to meeting household dietary needs. Temporary labour migration to Bangkok or rubber and coffee plantations in southern Thailand is increasingly important during the dry season, to alleviate population pressure and food scarcity. A reciprocal relationship is maintained whereby migrants are given rice and other resources from the village in return for remittances. Migration has facilitated the import of physical goods, modern values and knowledge to the village. These are challenging the traditional norms and institutions, and generating economic stratification.

The traditional agricultural system is reaching its limits in terms of production. Declines in forest cover and wild resources have been replaced by the cultivation of non-glutinous rice for market, but production cannot increase under current land and technological constraints. There is an increase in demand for consumer goods, importation of materials and hence cash through engagement with the market economy. While the village economy currently meets the community's needs, households would struggle to maintain food security if the price or demand for non-glutinous rice collapsed, or if the migrant labour market contracted. The agricultural system would also be unviable without labour currently provided by reciprocal workforce arrangements.

5

Remote, marginal and sustainable? The key role of brokers and bridging institutions for stronger Indigenous livelihoods in Australia's deserts

Jocelyn Davies, Yiheyis T. Maru, Fiona Walsh and Josie Douglas

The vast arid heart of Australia, remote from main centres of population and marginal to most industries, presents a distinct context and challenges for sustainability because of its variable climate, patchily distributed resources, sparse populations, very high proportion of Indigenous land and Indigenous populations and the gap in health status between Indigenous people and others. While desert Indigenous cultures co-evolved with the region's arid and variable climate and patchily distributed resources, characteristics of remoteness and marginality that define the region in the eyes of most other Australians are new.

Institutions crafted at national or state/territory levels often fail to engage Indigenous central Australians with opportunities that the region's social, economic and environmental context present for sustainable livelihoods. Conjoint poverty and rigidity traps characterise the region, as a result of mutual reinforcement between dense bonding networks and reciprocity among Indigenous people, and broader political and social environments that are often hostile to Indigenous values (Maru *et al.* 2012). Statistical inequality persists, masking a more fundamental issue for sustainability – social inequity, people's differential opportunities to lead lives that they have reason to value.

The sustainable livelihoods framework (SLF) has been valuable for understanding social dimensions of sustainability in this region because of its focus on the key role of institutions in determining people's opportunities to achieve outcomes that are meaningful to them. Brokers and bridging institutions have been found to be critical to addressing social inequities and promoting sustainability because of the big gulf between Indigenous socially embedded institutions, which emphasise relatedness as well as respecting individual autonomy, and the mechanisms through which governments engage with Indigenous people, predominantly as individuals, in efforts to address Indigenous social disadvantage. Our research illustrates the operation of brokers and bridging institutions in the intercultural space between Indigenous and bureaucratic settings in relation to Indigenous employment, commercial production of customary foods, in remote schools and in management of land for conservation outcomes and linkages to health outcomes.

An arid country

Australia has the distinction of being the continent with the highest proportion of land that is desert. Arid and semi-arid lands comprise 70% of the nation. Australia's rangelands,

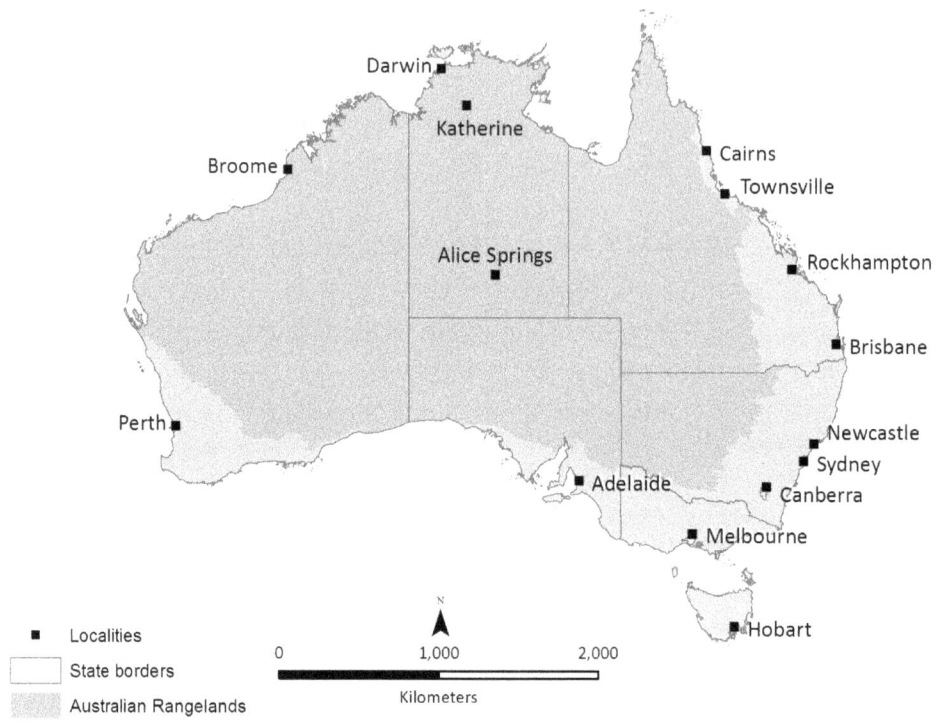

Fig. 5.1. The rangelands of Australia. Source: CSIRO.

comprising 80% of the country's land area, include these arid and semi-arid lands as well as the more northerly tropical savanna (Fig. 5.1). Known colloquially as the 'outback', Australia's arid, semi-arid and tropical savanna ecosystems comprise one of the last great natural areas on Earth (Woinarski *et al.* 2014) despite the considerable change to biota that has occurred in the past two centuries and that continues from changed fire regimes, feral herbivores and livestock, weeds and human-accelerated climate change. More than half a million people live in these lands (ABS 2012a; Ferguson 2012). They are commonly described as 'remote' because they are a long way from Australia's main population centres. These lands and their peoples do not conform to idealised images of Australian rurality – tamed, settled, transformed and economically productive (Prout and Howitt 2009). Rather, they are commonly represented as marginal to national agendas. The 21% of Indigenous Australians who live in the outback comprise close to 40% of the outback population. The relatively large proportion of the population who are Indigenous, together with the prevalent gap between Indigenous and other Australians in education, health and wealth, helps to account for outback regions being prominent among the areas of Australia where social disadvantage is concentrated (Vinson and Rawsthorne 2015).

Remoteness tends to be blamed for perpetuating, if not causing, dysfunction in Indigenous families and communities, for high rates of assault, imprisonment, suicide and child abuse. This view is poorly substantiated but resonates with policy-makers concerned about the difficulties and costs of providing police, education and health services in sparsely populated regions. Hence, an ongoing theme in political debate over the past two decades has been whether outback lifestyles, particularly those of Indigenous people, are sustainable and whether the places where outback people live are viable settlements (Stafford Smith et al. 2008; Prout and Howitt 2009; Taylor 2009). Calls to close remote Indigenous settlements arise as a result. These calls are countered by recognition that forcing people to move will have high social and financial costs as well as being unacceptable in terms of Australia's human rights standards (Taylor 2009). Sustaining the livelihoods of people who are motivated to live on or close to lands that they care about is also of central importance in maintaining the biodiversity, ecosystem services and natural and cultural heritage values of the continent (Woinarski et al. 2014). Over much of Australia's land mass, these people are Indigenous (Taylor 2011). Some evidence exists that the health of Indigenous people is better in very small outback settlements than in bigger centres that have more government-provided services (Rowley et al. 2008) and that the self-motivated engagement of Indigenous people in managing their customary outback lands has positive health outcomes and cost savings for health budgets (Burgess et al. 2009; Campbell et al. 2011; Davies et al. 2011).

These factors make it important to look beyond manifestations of social dysfunction in Indigenous communities and the economic cost of services when considering the sustainability of outback Australia. Structures and mechanisms that are less visible to casual observers enable Indigenous people to engage in productive activities in the outback, to cooperate and maintain identity, pride, independence and a sense of control over their own lives, and to steward their customary lands. In this chapter, we examine some of these structures and mechanisms through findings from empirical research and considerations from social-ecological systems theory that point to the important role of brokers and bridging organisations in desert Indigenous people's engagement with employment, education, markets and land management.

A central Australian perspective

Our perspective is from the middle of arid Australia, from the town of Alice Springs (population 28 500) and the central Australian region that it services. That region encompasses ~260 much smaller, dominantly Indigenous settlements with a total population of ~20 000 people distributed up to 800 km from the town (Taylor 2002; ABS 2012a). Census data, policy, and the architecture of government and community service provision promote a conceptual bifurcation of the region's population into two dichotomous groups, Indigenous and non-Indigenous. However, there is considerable diversity within each of these groups. For example, seven distinct languages and several more dialects are spoken by Indigenous people in Alice Springs and the region it services, representing nearly half of the remaining 'strong' Indigenous languages of Australia (Caffery 2010). Further, many people who identify as Indigenous also have forebears who were British, European, 'Afghan' or Chinese. The 'non-Indigenous' population is equally diverse, including fast-growing numbers of town residents from India, South Korea, China and various parts of the African continent as well as people of European heritage (ABS 2012b).

Alice Springs is categorised as remote in the standard statistical geography of Australia (ABS 2011). The hinterland that it services, together with most of the rest of rangeland

Australia, is categorised as very remote. In this geography, remoteness is defined by road distances between a place and towns or cities where various services are available. For the majority of Australians, this categorisation is unproblematic given that the region is so far from where they live. 'Remote' also invokes other perceptions among people who are distant to the region. They may perceive the region's landscapes to be wilderness or to have an 'otherness' – a stark beauty or fragility – and its people, particularly Indigenous people, to have an elusive spirituality and timelessness. Such perspectives add romance and the prospect of adventure to the nation's now dominantly urban narrative (Haynes 1998). At the same time, however, they can entrench the region in the national mindset as a relict whose contemporary purpose is obscure except perhaps where it is being mined. From that perspective, the outback might be like 'a colony from which natural resources are exploited and usually exported' (Woinarski et al. 2014). Indeed, the outback's mineral resources have generated some $90 billion annually in gross revenue to the nation in recent years (Ferguson 2012).

The perspective of Indigenous and other people whose identity is grounded in particular places in Australia's deserts, is invariably quite different. It sees these places and landscapes as home – the antithesis of remote – as central to identity and livelihood, and as resilient rather than fragile. However, since electronic control on how social security recipients spend their benefits has now extended the state's encounters with many Indigenous people into these intimate home spaces (Sider 2014), distance is not the barrier to enforcement of state social controls that it once was.

As with other regional service centres in desert Australia – Mt Isa, Kalgoorlie, Broken Hill, Port Hedland and Karratha being the larger ones – the mix of livelihoods in Alice Springs reflects regional resources and economic opportunities (Holmes 1997; Stafford Smith and Cribb 2009). Mining employment and service industries underpin the economy of most of those towns but are less significant in Alice Springs. Livelihood diversification and transformation has occurred progressively over the 150 years since non-Indigenous people first came to central Australia. In the early decades of colonisation, livelihoods were based solely on livestock grazing or on hunting and gathering foods – and on policing clashes between people engaged in these conflicting ways of life and their underlying resources, values, moralities and laws. Now the region offers a much wider array of market and amenity livelihood activities in a multifunctional landscape (Holmes 2002, 2010) with industries based on renewable natural resources and associated cultural resources (pastoralism, tourism, arts industries), industries that service the public interest (defence, conservation, public health, community services) and industries servicing other residents (small business, transport, government services).

Throughout these socio-economic transformations, outback livelihoods have been more uncertain than those in more densely populated regions as a result of an array of causally linked factors (Stafford Smith 2008). Relatively high social uncertainty is a result of distance from markets and centres of political decision-making. Sparse populations are inherently driven by the low productivity prevalent in rangeland environments. Very low and highly variable rainfall in deserts, and low soil fertility throughout the outback, account for this low productivity and make it inevitable that human populations will be low except in localities where they are sustained 'artificially' by water and food imported from elsewhere. Indigenous people's rights and cultural connections to land have survived better in these regions than other parts of Australia, because the low productivity meant that competition for land from other people has been later and less intense.

Net migration of both Indigenous and non-Indigenous populations is towards more urbanised areas – from very small settlements to regional centres and cities. Nevertheless, considerable numbers of people also move in the opposite direction (Biddle 2009). High job vacancy rates drive much of the movement of non-Indigenous people to the outback. Typical among non-Indigenous people is early career movement to the outback, for months or years, then shifting away later in life for career or family reasons, most often to a city (ABS 2012c). Shorter-term movement is also very common. For example, half the people who said Alice Springs was their usual place of residence at the time of the 2011 census said they lived somewhere else five years previously (ABS 2012b). Dominant policy and economic discourse constructs this as normal whereas Indigenous mobility, which tends to be even shorter-term and cyclic around a region, is constructed as deviant (Young and Doohan 1989; Prout and Howitt 2009). Mobility is an important strategy that desert people, as well as desert animals and plants, use to buffer the variability in the region – economic booms and busts as well as those driven very directly by rainfall (McAllister *et al.* 2009). Mobility also helps Indigenous people realise aspirations to maintain connections to family, country and culture. It strengthens intra-Indigenous networks and knits localities together. It may allow Indigenous people to maintain their independence and pride as a counterpoint to the prevalent rejection and exclusion from dominant society and its modes of communication and decision-making (Sider 2014).

As the characteristics described here suggest, two Australias – one dominantly Indigenous and one not – coexist in space and time in Alice Springs and some other outback areas to a much greater extent than in the east and south of Australia. A diverse array of brokers and bridging institutions are the mechanisms that enable these two Australias to cooperate and influence each other, and sometimes to flourish together. Anthropology has grappled with making this intercultural space analytically tractable, and in the process has come to appreciate Indigenous and non-Indigenous social forms not as separate domains, but as necessarily related to and influenced by each other (Hinkson and Smith 2005). Our own perspective draws also on structural analysis of social roles and networks (Scott 2012), in which flows of information/knowledge and other goods and services between people provide building blocks for understanding and describing their relationships, commonalities, differences and mutual influence.

We conclude that recognising and strengthening brokers and bridging institutions is critical for sustainability and resilience in the outback. In developing this conclusion, we first introduce the concepts of sustainability and resilience in relation to livelihood systems, and the sustainable livelihoods framework (SLF) (Fig. 5.2, case studies 1 and 2) which provides a conceptual model for the dynamic interactions that affect the quality of people's lives. We then illustrate the nature and efficacy of brokers and bridging institutions from four cases drawn from empirical research in central Australia: language and culture programs in bush schools (case study 2, from Douglas 2011), employment in the Anmatjere region (case study 3, from Davies and Maru 2010; Davies *et al.* 2010; Maru and Davies 2011), trading desert raisins (case study 4, from Holcombe *et al.* 2011; Walsh and Douglas 2011) and Indigenous employment in the Alice Springs Desert Park (case study 5, from Walsh and Davies 2011). We discuss the value and limitations of brokers and bridging institutions for sustainable livelihoods and sustainable development in desert Australia.

Sustainability and resilience in livelihood systems

'Sustainability' and 'sustainable development' are sometimes used interchangeably. However, sustainability is better considered as a property of a system: the system's ability to continue through time as it maintains critical resources (often natural resources), institutes coping strategies in the face of shocks and stresses, and adjusts these through adaptive learning processes (Maru and Woodford 2007). Sustainable development is a normative concept that incorporates ethical considerations of fairness and equity. Through sustainable development, all people should have fair and equitable opportunity to realise 'capability' as a livelihood outcome provided they do not gain such outcomes at the cost of other people, including future generations.

Very many people from all cultures live safe, productive and fulfilled lives in the outback. For many Indigenous people, fulfilment involves considerable investment of time and personal resources to maintain cultural strength and attachment to their traditional country even though the dominant political and economic climate may be quite hostile to their efforts. Nevertheless, the current social situation is a long way from the ideal of sustainable development. Social indicators show very high rates of Indigenous morbidity and of arrest, imprisonment and violent assault that most often involves Indigenous people as both victims and perpetrators. Various forms of inequality – in income, wealth, health, education – are concerns that Australians commonly express when considering the nation's future, because they reduce social cohesion (Cork et al. 2015). Such inequalities are correlated (Kondo et al. 2009) and are high or extreme in many of the outback regions that have high Indigenous populations (SCRGSP 2014; Fleming and Measham 2015; Vinson and Rawsthorne 2015). These inequalities distort private and public investments in social and economic infrastructure towards policing social order, providing emergency medical treatment and punishing truancy, at the expense of building community, assuring public health and celebrating learning. Marginalisation of Indigenous people perpetuates their social and economic disadvantage and is the antithesis of sustainable development (Berry 1997).

Resilience concepts add a deeper dimension to consideration of the dynamics and sustainability of livelihood systems. Resilience is defined as 'the capacity of a social-ecological system to absorb disturbance and reorganise while undergoing change so as to still retain essentially the same function, structure and feedbacks – and therefore the same identity' (Resilience Alliance 2002). Because it confers resistance to external shocks, and flexibility to recover from them, resilience is very often assumed to be normative – a 'good thing' that all systems, people, communities and corporations should strive for. However, stable but unhealthy states may also have high resilience (Walker et al. 2004), such as when individuals and families are trapped in poverty and vulnerability. This most undesirable situation is more likely to persist – that is, to be resilient – where individuals are very highly connected socially to others in the same situation and have few other social connections (Maru et al. 2012). Such individuals may have abundant bonding social capital, or supportive connections to their own family and peers, but lack bridging and linking social capital, or linkages to people with different social, cultural and economic characteristics (Woolcock and Narayan 2000). They will typically lack connections across levels of governance or decision-making, and so lack capacity to influence people outside their own family or friendship circle (Armitage 2007). The inward-looking networks of Indigenous minorities that characterise such situations are reinforced when juxtaposed with institutions of the dominant culture and state that are alienating and rigidly imposed (Maru et al. 2012).

The SLF encourages a systems understanding of how people's livelihoods are shaped by their own values and assets, by those of other people, by the external environment and by institutions enacted through regularised patterns of social behaviour. Our use of this framework in research for development in central Australia has built on international experience (Davies et al. 2008). Through dynamic interactions and feedbacks among various elements of the framework (Fig. 5.2, case study 1), people use the capitals that they can access (case study 2) in ways that prevailing institutions allow, to generate outcomes for their livelihoods. Livelihood outcomes can be conceived of in material terms, such as income, food and shelter. They also encompass higher-order conceptions of what people see as important for a 'good life'. This dimension of the SLF interfaces with many other efforts to define the dimensions of human well-being or happiness. Amartya Sen's work has been influential. Sen developed the concept of 'capability', which he defined as the ability of human beings to lead lives they have reason to value and to enhance the substantive choices they have (Sen 1999). 'Capability' is a social goal, applicable to all Australians. For example, it provides a conceptual underpinning to the Australian Treasury's well-being framework (Gorecki and Kelly 2012). Diminished capability, such as through the disempowerment and loss of control over their own lives that desert Indigenous people commonly experience, reduces people's capacity to buffer the impact of other chronic stressors, lowering resistance to disease and increasing morbidity (Ursin and Eriksen 2010; Davies *et al.* 2011).

Equity, stemming from fair distribution of assets, risk and influence, is an integral part of the concept of sustainable livelihoods. However, prevailing public institutions tend to be ineffective at managing risks to the livelihood assets of marginalised people. By definition, such people have had little influence on the design or implementation of mainstream institutions, which makes conflict with their own socially embedded institutions highly likely. As the interactions portrayed in the SLF indicate (Fig. 5.2), marginalised people gain increased influence over dominant institutions when their assets are more highly valued by other people. Brokers and bridging institutions enable that increase in influence.

Case study 1: Sustainable livelihoods framework

The various components of the SLF (Fig. 5.2) interact to influence the outcomes that people experience in their lives. These components are as follows.

- Livelihood strategies or activities: what people actually do in their lives that leads to outcomes.
  Our specific examples (case studies 3–6) encompass waged employment through teaching in small 'bush' schools, in horticultural enterprises, community services and a conservation-based tourism enterprise, and activities of harvesting and trading bush foods.
- Livelihood assets, or capitals, which may be conceptualised as human, social, natural, physical and financial.
  The capitals that enable Indigenous people in central Australia to engage in commercial harvest of bush foods (case study 2) particularly exemplify the roles of traditional knowledge and relationship networks in enabling contemporary

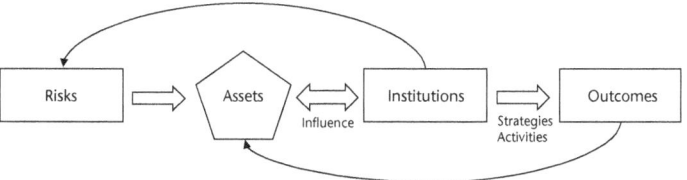

Fig. 5.2. SLF showing generic components and relationships. Source: Modified from Davies *et al.* (2008).

Indigenous livelihood activities. The dispersed pattern of small settlements in central Australia and the linking road network are physical capitals that are key enablers of commercial bush harvesting and many other contemporary Indigenous desert livelihood activities.

- Institutions, or rules and norms (Ostrom 2005), which may be conscious or tacit.
  Feedback arrows and interactions in the framework (Fig. 5.2) indicate that the social role of institutions is to manage risk to people's lives. Institutions may stem from people's own culture and customs or be formally crafted, collectively by groups of people, or by governments or corporations. Laws and rules that have been overtly agreed are specific institutional forms, often being 'public' institutions in that they aim to keep good order in society and prevent or resolve conflicts between people. More broadly, institutions can be thought of as modes of behaviour or ways of doing things that are socially embedded, accepted or expected ways of operating within a social group. If people do things differently from such accepted modes of behaviour they may be considered by others in their social group to be behaving oddly, if not dangerously or antisocially.
- Risks, or the vulnerability context, which stem from the external environment in which people live.
  Risks include trends and shocks that will impact on people's assets and their capacity to use those assets in generating outcomes for their lives. Examples in desert Australia are warming trends under climate change (Healy 2015), intense rainfall that generates erosion and flooding, policy change affecting local employment opportunities or essential service provision, and changing market preferences affecting demand for produce. Uncertainty caused by the high temporal and spatial variability of desert rainfall adds a further dynamic with pervasive impact: desert people engage more readily with new ways of doing things when it has rained, than when it has not (Robinson et al. 2009; Measham et al. 2011).

Case study 2: Capitals that commonly enable desert Indigenous livelihood activities

Derived from Holcombe et al. (2011); Walsh and Douglas (2011).

About 5% of the Indigenous people who live in 25 small central Australian settlements have engaged in commercial harvest of seeds and fruits from native plants over ~15 years. The assets or capitals that have enabled these 400 people, mainly middle-aged and older women, to develop commercial harvest as a livelihood activity are described below. This description exemplifies the ongoing value of traditional knowledge and relationship networks in this and other contemporary Indigenous livelihood activities, together with the value of 'modern' physical and financial infrastructure.

Human capital and intellectual capital

Harvesters invest their labour, applying techniques based on traditional practices. They also apply:

- ecological knowledge of plants to species level, including local names, identification of harvested portions and safe use of plants, including species that are not safe to trade;
- spatial and temporal knowledge of species' occurrence and productive periods;

- skills and labour to manage production through burning productive areas and through ceremony;
- negotiation skills in engaging with traditional landowners to secure permission to harvest and in establishing sale prices with traders;
- knowledge, skills and labour to manually harvest from many plants across wide areas then sort, clean, dry and store products and be competent in other necessary tasks.

Natural capital
- Entitlement to access land for harvest.
- Biodiversity, which facilitates switching target species depending on seasonal conditions and market demand.
- Rainfall and fire regimes amenable to production.

Physical capital
- Remote settlement pattern and road network (facilitating access for ready monitoring of ripeness and for harvest operations).
- Access to family vehicle and driver (men usually do the driving).
- Access to telephone (to contact traders about market demand).
- Limited but robust equipment for picking, drying and short-term storage.

Social capital
- Relationships with one or more traders.
- Family cooperation.
- Relationships with neighbouring harvesters.
- Relationships with traditional landowners.

Financial capital
Other income sources to sustain harvesters during non-harvest periods and enable pre-payment of harvest costs such as vehicle fuel:
- social security;
- waged employment (12% of harvesters);
- income from art sales, and mining royalty payments to traditional landowners.

Case study 3: Trading desert raisins
Source: Walsh and Douglas (2011).

Small-scale commercial harvest of Indigenous plant foods has taken place for three decades in central Australia. This small and largely invisible economy has been enabled by harvesters having access to necessary capitals (case study 2) and by a small number of traders who have brokered between harvesters and markets. Indigenous harvesters have gained information about market demand, which has been strongest for the fruits of the desert raisin (also known as the bush tomato, *Solanum centrale*), only through their relationships with traders. Likewise, traders have been able to operate because of their sustained and trusted relations with harvesters.

Traders' needs for working financial capital are high relative to the volume of produce traded because their suppliers, the Indigenous harvesters, work to a cash economy and require payment when their harvests are transferred to the trader. By contrast, the manufacturers and retailers who are the traders' customers pay on invoice after a shipment is delivered. Traders need to bridge the gap between the relatively small but steady demand from companies that manufacture and retail products containing desert raisin, and extreme fluctuations in supply due to the variability of rainfall and amount of harvested produce. These fluctuations meant that one trader could purchase 3 tonnes in 2012 whereas a median annual purchase over a 12-year period was <100 kg, inadequate to meet demand. Traders have needed to develop their physical capital in the form of storage facilities for years when supply is much greater than demand.

Ways of operating (institutions) that traders have developed to engage with Indigenous harvesters include travelling long distances to Indigenous settlements when seasonal conditions indicate good harvests are likely, communicating about how much produce they plan to purchase by distributing the corresponding number of sacks among interested harvesters (taking into account harvesters' assessments of anticipated production volumes), and conducting weighing, quality assessment of harvests and cash payment to harvesters during buying trips 'publicly' (with other harvesters present) to promote transparency in transactions.

The high costs that traders incur from these institutions have been made more manageable by each trader operating within only one part of the central Australian region. The traders' track record of 'doing what they say they will do' has built long-term trust relationships with harvesters – social capital that has been critical to trader livelihoods in this high-risk business environment. Because financial returns from bush harvesting and from trading are not high and are highly uncertain, harvesters and traders are also involved in other livelihood activities and have other sources of income.

Case study 4: Indigenous Language and Culture programs in bush schools

Source: Douglas (2011).

Engagement of families with their children in schools promotes educational outcomes for children. Learning strategies that build on community aspirations for education help create connections between home and school, linking families and schools as learning communities (Wenger 1999; Schwab and Sutherland 2003). A rich picture of how these outcomes are generated in schools in small Indigenous settlements in central Australia was developed through qualitative research in 2007–08 in two communities whose schools had incorporated Indigenous Language and Culture (ILC) programs into their teaching and learning. This picture contrasts markedly with prevalent portrayals of failure in Indigenous education.

From 2006, the Northern Territory Curriculum Framework (NTCF) included an elective ILC component. This was the first formal policy recognition of Indigenous language and culture in Northern Territory education other than in designated bilingual schools. The NTCF provides for language maintenance and language revitalisation, and for cultural teaching and learning activities that can be founded in local languages and practices (NTDET 2011). Content outcomes are based on Indigenous organisation of knowledge, such as 'Country/land' and 'People and Kinship', and link to other curriculum learning areas. The scope, focus and delivery of ILC programs varies between schools. The ILC programs at Ntaria and Willowra, the two schools involved in this research, used experiential practice-based learning outside the classroom involving country visits – day excursions or overnight camps on country of significance to the school's community.

Inclusion of local language and cultures in the teaching/learning programs of desert Indigenous communities is a bridging institution that enables family and broader community involvement in the school. Indigenous teaching assistants, from the schools' local communities, are the brokers who enable this bridge. They place high value on working collaboratively and in partnership with non-Indigenous teachers. In these schools, engagement of children with any school-based learning relies on the Indigenous teaching assistants – students often speak little English at home and the non-Indigenous teachers, who comprise the majority of school staff, rarely have knowledge of local languages. Classroom teaching of a standard curriculum provided few opportunities for the Indigenous teaching assistants to take leadership roles. In contrast, ILC programs provided a ready opportunity. Indeed, implementation of ILC programs depended on the human and social capital of Indigenous teaching assistants.

The country visits that were central to ILC programs involved school children, Indigenous and non-Indigenous staff from the school, elders and other family members sometimes including teenagers who are no longer attending school. The experiential and practice-based approach to teaching and learning during country visits reflected customary Indigenous pedagogies. This enabled intergenerational transfer of knowledge that is context and site-specific.

The key role of Indigenous teaching assistants started with them engaging community elders in decisions about what area of country should be visited. Together they ensured the location would be appropriate to curriculum learning themes and to the knowledge they considered the students needed, and they organised how the visit would take place. Indigenous teaching assistants produced teaching/learning resources relevant to the visit such as worksheets, flashcards and books – literacy resources that built from the students' experiences of the trip. These books were very highly valued in the schools.

ILC programs helped to align local educational goals and values with those of the broader educational system and laid a foundation for strong relationships of mutual trust and respect. They provided the opportunity for Indigenous teaching assistants to have their ideas valued, respected and acted upon in the school, building their confidence and morale. This leadership role brought balance to their partnerships with non-Indigenous teachers. Community engagement was enhanced because relationships between community and school depend on the quality of relationships between Indigenous and non-Indigenous teaching staff.

Case study 5: Jobs in Anmatjere region

Sources: Davies and Maru (2010); Davies *et al.* (2010); Maru and Davies (2011).

In the Anmatjere region, 200 km north of Alice Springs, the conundrum of 'lots of jobs, lots of unemployed Indigenous people' was identified by local leaders as a core issue for sustainable development. Horticulture, rangeland beef production, government, community services and retail are the region's main industries, and mining is proposed. However, few Indigenous people outside the leadership group shared the perception that 'lots of jobs' existed, indicating that many of the employment opportunities were physically or socially inaccessible to them.

Two sets of norms – from Indigenous culture and what we term 'workplace culture' – meet daily in the Anmatjere region. 'Workplace culture' includes norms for how to behave in a workplace, where interactions among people are based on roles rather than social relationships. Private sector employers in the Anmatjere region, all of them non-

Table 5.1. Workplace norms implied by private sector employees' comments about employment of local Indigenous people

Comments from private-sector employer interviewees about employment of local Indigenous people	Implied workplace norm (after Crawford and Ostrom 2005)
Indigenous people will not come to work alone, they need to be in a group	A worker must be able to operate autonomously in the workplace
Poor functional literacy and numeracy is a barrier to employment	A worker must be able to operate autonomously in the workplace, including tasks that require functional literacy/numeracy
Indigenous people need retraining every day and won't take the initiative	A worker must take initiative in relation to their work responsibilities
Lack of endurance/capacity to sustain and complete the hard work required in the horticulture industry	A worker must work a full shift
Indigenous people can only be expected to work from 8 a.m. to 12 p.m., they are unable to work long hours for a variety of reasons	A worker must work a full shift
Lack of desire or motivation to work	A worker must be self-motivated
In employing Indigenous people, the employer may have to spend extra time giving support and 'hassling' employees to get them to attend punctually	A worker must be self-motivated, self-reliant, organised and punctual

Indigenous, implied that several such norms were factors that made it difficult for them to employ local Indigenous people (Table 5.1). These unspoken requirements of workplaces are common-sense to those who know them and daunting to those who do not.

Standard ways for desert Indigenous people to learn workplace culture include work-ready training and traineeships. However, Indigenous people from the Anmatjere region rarely mentioned these as a way to get work. Rather, they indicated that social capital through kin and other trusted relationships was very important. A relatively small number of individuals, both Indigenous and non-Indigenous, were helping others navigate the uncertainties of engaging with workplace culture. These unpaid brokers assisted with phone calls, documentation, introductions or knowledge that other people required to access work. The operation of social capital was also apparent, from Indigenous people who described being 'picked' or 'invited' into employment and a willingness expressed by employed people to facilitate entry into work opportunities for others. A strong employment record by particular families in some workplaces indicated the outcome of these strategies.

Employers in the region vary in the extent to which they adjust employment conditions to better match Indigenous norms. Indigenous people considered some government and community work environments to be good because they had procedures for Indigenous people to get leave where family or cultural responsibilities required them to be absent from work. In other sectors, notably horticulture, the tight timeframes of seasons and markets made the competing demands on the time of Indigenous employees from culture and family very risky to employers. One Indigenous organisation contracted for horticultural work, engaging work groups of local Indigenous people as one way to spread this risk. Shared norms that developed within the work groups also buffered the impact of the clash between Indigenous cultural and workplace norms. Where effective brokerage and bridging mechanisms do not exist, such clashes can readily reduce Indigenous people's motivation for employment. Rather than them developing loyalty to employer and work associates, their loyalties to family, local community and cultural responsibilities are strengthened.

Case study 6: Indigenous employment in the Alice Springs Desert Park

Source: Walsh and Davies (2011).

Alice Springs Desert Park, an enterprise owned and operated by the Northern Territory government, has been considered by many observers as a best practice model for Indigenous employment compared to other government workplaces and many other organisations. Business, political and social justice imperatives drove the park's managers to develop its Indigenous employment policy. That policy has operated as a bridging institution to underpin the park's success in Indigenous employment. Recognition by non-Indigenous employees of the constant negativity and racism that Indigenous people experienced in their lives and work in Alice Springs contributed to their support for the policy and its implementation.

Many of the Indigenous employment processes at the park were readily comparable to those recommended in an Indigenous engagement and evaluation tool developed for mining companies and mine sites (Tiplady and Barclay 2007). That tool established a pathway to Indigenous employment outcomes through corporate processes and accountabilities of leadership, planning, implementation, monitoring performance and assessing progress. At the Desert Park, two additional important pathways to effective Indigenous employment were apparent.

First, the park's objectives, master plan and development plan, as well as its Indigenous employment objectives, established that Indigenous staff (and only Indigenous staff) would interpret and present Indigenous culture to visitors. Second, Indigenous people filled multiple roles in the park. While some Indigenous people were full-time park employees, others were involved in paid or voluntary roles as cultural experts, mentors and artists. Involvement of young Indigenous people was promoted through the park's 'Adopt a School' program. Work teams from the local prison, mostly comprising Indigenous people, also worked in the park. Park staff maintained good relationships with traditional Indigenous owners of other localities in central Australia that are connected to the park through customary custodial networks or are localities where park staff have collected biological specimens.

Key individuals, both Indigenous and non-Indigenous, who had sustained a long-term commitment to keeping Indigenous employment strong, brokered between Arrernte and workplace norms. Park managers overtly recognised the park as Arrernte customary lands, according traditional owners a right of veto over decisions in the park even though the Federal Court had found that native title to the park area had been extinguished. This recognition of Arrernte customary authority demonstrated consistency with deeply entrenched Indigenous cultural protocols that were important to Indigenous employees. It established a firm foundation for Indigenous employees to feel comfortable and develop confidence in the work environment of the park.

Indigenous and non-Indigenous staff had similar views about some of the institutions that they considered were important to the effectiveness of Indigenous employment in the park. Both groups identified the traditional owners' right of veto, the interpretation of Indigenous culture in the park only by Indigenous people, the good relationships between Indigenous and non-Indigenous staff, the commitment and action on Indigenous employment by key individuals, and regular interactions between park employees and Indigenous people involved in diverse roles in the park's operations. The park's Indigenous staff also noted the importance of the opportunities that their work provided to learn more about Indigenous culture and to interpret not only culture but the natural environment to visitors, the diversity of tasks, the outdoor work location, and the application of Indigenous cultural modes of organisation in the workplace in that respect was accorded not just to role, rank and expertise but also on the basis of age and affiliation to traditional owners. Indigenous employees drew particular attention to the need to maintain and strengthen these elements in order to sustain the park's strong track record in Indigenous employment.

Brokers and bridging institutions

Brokers are people who link across structural holes in social networks. Where people are closely linked to other people in a very tightly bonded group, they tend to have relatively few links to people outside that group. This lack of links is termed a structural hole (Burt 2005). Information and ideas move quickly among members of a tightly bonded group, and trust is high, because people in the group know what to expect from each other. They adhere to the same norms or ways of behaving, the same institutions. However, new ideas or information (human and intellectual capital, case study 2) are not readily transmitted to the group because there is a structural hole, and the group's ideas and information are not readily accessible to people outside the group. As people who link across structural holes, brokers – by definition – have access to the group's ideas and information and those of other people the broker links to.

The broker concept is familiar due to the roles played by brokers in markets, such as in sourcing insurance and mortgages for individual clients. In a more general sense, brokers may play an open role in communication between groups, transmitting ideas and information or fostering more direct relationships and trust among individuals in different groups. Alternatively, they may carefully manage such flows. They can be expected to strategically use their access to both groups to advantage themselves in some way, whether materially or in terms of reputation. Altruism, stemming from a desire to improve the situation for one group and/or the other, may be a motivation but is not likely in itself to sustain brokers in the high-stress role of linking between Indigenous and non-Indigenous groups in the rangelands (Maru and Davies 2011).

Empirical research on the livelihoods of Indigenous and non-Indigenous people in Anmatjere region (case study 5) revealed that employed Indigenous people had most often secured their jobs with the help of relatives or other members of their community who acted as brokers. This kind of brokerage was the pathway that local people recommended as the most effective way for other community members to get employment. These brokers were not very visible and the pathways they facilitated were little recognised by governments, who directed their investments into skilling Indigenous people in more standard job search and training mechanisms. Similarly, in the bush food industry (case study 3), a small number of traders brokered across the structural hole between the networks of Indigenous harvesters and those of the food processors in eastern and southern Australia who use bush raisins in chutneys and sauces. In bush schools (case study 4), which are the only way for most children in small outback Indigenous communities to acquire literacy,

Indigenous teaching assistants are key brokers between the school's non-Indigenous teachers and the school's local community. Their effectiveness was found to be enhanced in schools that implemented ILC programs, which serve as bridging institutions.

Bridging institutions are ways of doing things that respect and accommodate the different sets of norms and accountabilities that characterise different social or cultural groups. The numerous protocols and guides that advise how non-Indigenous people should behave in Indigenous settings indicate the extent of difference between the norms and accountabilities of Indigenous and other social groups. Such protocols and guides are themselves bridging institutions. 'Two-way' and 'both ways' are terms in common use in Indigenous land management (Muller 2012; Preuss and Dixon 2012) and in education (Harris 1990; White *et al.* 2009; Ngurruwutthun 2014), that encapsulate the concept of bridging institutions. They refer to communication, teaching and learning practices that are effective in meeting the aspirations of Indigenous people to do things in ways that have high standing and relevance in both their own and non-Indigenous cultures, that draw on Indigenous knowledge and other knowledge systems, and that equip young people for navigating between cultures.

In bush schools, ILC programs can bridge between the community's goals for education and mainstream curriculum requirements (case study 4). Bridging institutions that were developed to span the differences in norms between Indigenous people and workplaces in Anmatjere region (case study 5) included the formation of work groups of Indigenous people contracted through an Indigenous organisation. In desert raisin trade (case study 3), traders developed specialised ways of operating to make effective bridges between their dispersed supply network, the extreme fluctuations in desert raisin supply that are driven by variable rainfall, and market demand. The Indigenous employment policy of the Alice Springs Desert Park has been an important bridging institution, effective in fostering a strong track record of Indigenous employment (case study 6). The bridging institutions that underpin that success include park management's recognition of Arrernte customary land ownership.

Outcomes for capability are strong when brokers and bridging institutions are operating effectively. In bush schools, Indigenous teaching assistants have a rare leadership role in ILC programs (case study 4). This helps to bring balance to the teaching assistants' relationships with non-Indigenous teachers, facilitate mutual trust and respect, promote engagement between families and the school and encourage families to value school attendance. Effective bridging institutions have resulted in the Alice Springs Desert Park gaining a reputation for best practice in Indigenous employment and this has generated confident and valued Indigenous staff (case study 6). Capability outcomes from desert raisin trading (case study 3) are also high, notably in the pride that harvesters show in purposeful self-managed work that keeps traditional ecological knowledge, skills and governance vital and relevant, and that is independent from and not reliant on any government agency or other organisation.

Frustrations with brokers and bridging institutions, fragilities and failures, are also apparent. In Anmatjere region (case study 5), employment brokering has to date not succeeded in engaging Indigenous people with economic opportunities in horticulture. While bridging institutions such as work groups could manage the risk of discontinuities in the attendance of individual Indigenous people at horticultural work sites, they could not overcome the low motivation of most Indigenous people towards horticultural piecework. This is due, at least in part, to the horticultural crops and their management having no cultural significance (Rea and Messner 2008) unlike desert raisins collected by wild harvest (case study 3).

Decisions about whether ILC programs are offered in bush schools (case study 4) are part of the school principal's responsibility. Systemic outcomes from those programs for community trust of, and engagement with, the school may not be readily appreciated by principals. Requirements to give primacy to rapidly improving student literacy may be seen as conflicting. High turnover of non-Indigenous staff in remote schools also obviates measures to build on success or address weaknesses of past efforts. At the Alice Springs Desert Park (case study 6), Indigenous employment success has remained central to the park's strategic goals even with changes in management. However, key elements of the bridging institutions that enabled that success are not readily transferable to businesses that are not concerned with natural and cultural resource management.

Analysis of desert raisin trading using a standard agri-food value chain approach (Bryceson 2008) found that the trust and good communication apparent between traders and harvesters (case study 3) did not exist throughout the value chain, and indicated that the traders' role as broker was not universally valued. Desert raisin harvesters' reliance on traders for market information has been seen as an industry weakness (Cleary 2012). Stronger and more direct linkages between harvesters and processors have been advocated (Cleary 2012). Certainly, reliance on a single broker can make Indigenous people's information networks vulnerable (Woodward 2008). However, the physical and social distance between harvesters and city-based processors, and associated communication challenges, suggest that sustaining information flows and trust will be difficult without people in broker roles and raises the question of how those people would gain a livelihood if not from trading. Research into desert raisin horticulture, improved cultivars and irrigation regimes has sought to overcome discontinuity in desert raisin market supplies due to rainfall variability, while maintaining Indigenous involvement in and benefit from the industry (Lee 2012). Optimising production is a standard agricultural approach but has trade-offs for resilience (Walker and Salt 2006). Horticultural production, being divorced from other capitals that Indigenous people invest in when harvesting from the bush (case study 2), can readily obviate outcomes for maintenance of harvesters' place-based identities, language and customary governance and can reduce harvesters' sense of control over their lives. This perverse outcome was indicated by the reaction of a group of harvesters who were shown irrigated fields of desert raisins during a drought when there was no production 'out bush'. They concluded that non-Indigenous people had taken all the desert raisins for their own use and financial benefit (Cleary 2012).

These examples indicate that brokering and bridging institutions that are developed at small scale can be vulnerable or inadequate when efforts are made to extend their scope and scale. However, the Indigenous land and sea management industry provides a strong example of successful scaling. This industry started from the strong aspirations and agency of outback Indigenous people and has grown rapidly over the past two decades, delivering strongly to both Indigenous and dominant society goals (Hill et al. 2013). The industry has core support from two Indigenous-specific Australian government programs: the Indigenous Protected Area program, aimed at conservation of biodiversity, Indigenous ecological knowledge and cultural heritage, and the Working on Country program (Hill et al. 2013; Smyth 2011) aimed at building Indigenous employment and effective natural and cultural resource management (AAG 2011; Smyth 2011; Davies et al. 2013; Hill et al. 2013). Success factors in these programs include that they are implemented flexibly by governments and provide for brokers to have strong roles. Many of the brokers in Indigenous land and sea management are staff of Indigenous regional representative organisations and regional natural resource management bodies that have invested in building networks with their Indigenous constituents (Hill *et al.* 2013). Land management professionals have developed understandings about effective brokering and bridging institutions through fieldwork with Indigenous people and have encouraged their peers to emulate those practices (Horstman and Wightman 2001; Hoffmann *et al.* 2012; Robinson and Wallington 2012). Indigenous rangers funded through the Working on Country program and other sources are developing key roles as brokers, linking between members of their own community and outsiders with science-based knowledge relevant to land management.

A growing track record of outcomes for Indigenous people's capability and for natural and cultural resource condition is emerging from effective brokering and bridging institutions, including economic mitigation of climate change with social and ecological benefits (Russell-Smith *et al.* 2013; Robinson *et al.* 2014), new funding and partnership models for landscape-scale management (Salmon and Gerritsen 2013; Jupp *et al.* 2015), and new land management networks linking across the vastness of Australia's deserts (Ten Deserts Initiative 2015). Environmental outcomes from such efforts appear to be fostering a strong willingness in mainstream Australian society to pay for Indigenous land and sea management (Zander 2013).

Conclusion: building sustainable and resilient livelihoods

Social dysfunction and the high cost of delivering services to sparse and dispersed populations provide justification for imposing 'modern' ways of doing things and systems of value onto outback Indigenous people. The assumption with 'modernisation' is that 'local ties and parochial perspectives [will] give way to universal commitments and cosmopolitan attitudes; that the truths of utility, calculation and science [will] take precedence over those of the emotions, the sacred and the non-rational; [and] that the individual rather than the group [will] be the primary unit of society and politics' (Rudolph and Hoeber Rudolph 1967, pp. 3–4). Consistent with this trajectory, 'real' decisions that affect desert Indigenous people are made 'by policy makers and politicians, often with little effective communication with people on the ground' (Ferguson 2012, p. 28). Problems are frequently encountered in implementation, for example due to the paucity of mechanisms for effective intra- and inter-agency communication and collaboration, absence of flexible funding, and confusion about priorities or starting points for action (Ferguson 2012; Walker *et al.* 2012). Ineffective ways of operating can readily become entrenched because bureaucratic imagination and political will have limited capacity to develop and implement alternatives (Lea 2008). Although Indigenous people's remoteness and attachment to traditional values and behaviours are often blamed for the persistence of undesirable livelihood outcomes, blindness and rigidity in dominant institutions are equally culpable (Maru *et al.* 2012). Overriding what people value, even with benevolent intentions, threatens their identity and generates mistrust. This can reinforce insularity in people's networks and maintain resistance to change in social systems even when the outcomes thus perpetuated are destructive to people's well-being and exacerbate social inequities.

A livelihoods perspective encourages attention to the assets and ways of operating that people actually value and use. In this chapter we have drawn on examples of livelihood strategies that show the importance of brokers and bridging institutions for generating positive livelihood outcomes in desert Australia. Brokers and bridging institutions face significant challenges including lack of recognition, stress from conflicting and mismatching institutions, and rapid changes in policy or political attention. None of the case studies offer a panacea for the challenges of building sustainable livelihoods. However, they highlight the importance of not overlooking or overriding assets and ways of operating that Indigenous and other outback people have and value, and that are generating positive outcomes for capability. Rather, it is important to be attentive to opportunities to strengthen these assets and institutions, and to develop strategies and partnerships that generate trust by engaging and expanding them.

References

AAG (Australian Auditor-General) (2011) *Indigenous Protected Areas*. Audit Report No. 14 2011-12 Performance Audit. Australian National Audit Office, Canberra.

ABS (Australian Bureau of Statistics) (2011) *Australian Statistical Geography Standard: Volume 5 – Remoteness Structure*. Cat. 1270.0.55.005. ABS, Canberra.

ABS (Australian Bureau of Statistics) (2012a) *2011 Census of Population and Housing*. ABS, Canberra.

ABS (Australian Bureau of Statistics) (2012b) *2011 Census of Population and Housing: Basic Community Profile Alice Springs (T) (LGA70200), based on place of usual residence*. Cat. 2001.0. ABS, Canberra.

ABS (Australian Bureau of Statistics) (2012c) *Still on the Move. Reflecting a Nation – Stories from the 2011 Census, 2012-2013*. Cat. 2071.0. ABS, Canberra.

Armitage D (2007) Building resilient livelihoods through adaptive co-management: the role of adaptive capacity. In *Adaptive Co-management: Collaboration, Learning and Multi-level Governance*. (Eds D Armitage, F Berkes, NC Doubleday) pp. 62–82. UBC Press, Vancouver.

Berry JW (1997) Immigration, acculturation, and adaptation. *Applied Psychology* **46**(1), 5–34.

Biddle N (2009) *The Geography and Demography of Indigenous Migration: Insights for Policy and Planning*. Working Paper No. 58/2009. Centre for Aboriginal Economic Policy Research, Australian National University, Canberra.

Bryceson KP (2008) *Value Chain Analysis of Bush Tomato and Wattle Seed Products*. Research Report 40. Desert Knowledge CRC, Alice Springs. <http://www.nintione.com.au/resource/DKCRC-Report-40-Value-Chain-Analysis.pdf>.

Burgess CP, Johnston FH, Berry HL, McDonnell J, Yibarbuk D, Gunabarra C, Mileran A, Bailie RS (2009) Healthy country, healthy people: the relationship between Indigenous health status and 'caring for country'. *Medical Journal of Australia* **190**(10), 567–572.

Burt R (2005) *Brokerage and Closure: An Introduction to Social Capital*. Oxford University Press, New York.

Caffery J (2010) Central Australian endangered languages: so what? *Dialogue: Journal of the Academy of Social Sciences in Australia* **29**(1), 78–87.

Campbell D, Burgess P, Garnett ST, Wakerman J (2011) Potential primary health care savings for chronic disease associated with Australian Aboriginal involvement in land management. *Health Policy* **99**(1), 83–89. doi:10.1016/j.healthpol.2010.07.009

Cleary J (2012) It would be good to know where our food goes: information equals power. In *Indigenous Peoples' Innovation: Intellectual Property Pathways to Development*. (Eds P Drahos, S Frankel) pp. 8–10. ANU e-Press, Canberra.

Cork S, Grigg N, Alford K, Finnigan J, Fulton B, Raupach M (2015) *Australia 2050: Structuring Conversations about our Future*. Australian Academy of Sciences, Canberra.

Crawford E, Ostrom E (2005) A grammar of institutions. In *Understanding Institutional Diversity*. (Ed. E Ostrom) pp. 137–173. Princeton University Press, Princeton.

Davies J, Maru Y (2010) Living to work or working to live: intercultural understandings of livelihoods. *Dialogue: Journal of the Academy of Social Sciences in Australia* **29**(1), 18–34.

Davies J, White J, Wright A, Maru Y, LaFlamme M (2008) Applying the sustainable livelihoods approach in Australian desert Aboriginal development. *Rangeland Journal* **30**(1), 55–65. doi:10.1071/RJ07038

Davies J, Maru Y, Hueneke H, Grey-Gardner R, Chewings V (2010) *Outback Livelihoods: Employment, Sustainable Livelihoods and Development in Anmatjere Region, Central Australia*. Ninti One Ltd, Alice Springs. <http://www.nintione.com.au/resource/dkcrc-report-61-outback-livelihoods_employment-sustainable-livelihoods-and-development-in-anmatjere-region-central-australia.pdf>.

Davies J, Campbell D, Campbell M, Douglas J, Hueneke H, LaFlamme M, Pearson D, Preuss K, Walker J, Walsh FJ (2011) Attention to four key principles can promote health outcomes from desert Aboriginal land management. *Rangeland Journal* **33**(4), 417–431. doi:10.1071/RJ11031

Davies J, Hill R, Sandford M, Walsh F, Smyth D, Holmes M (2013) Innovation in management plans for Community Conserved Areas: experiences from Australian Indigenous Protected Areas. *Ecology and Society* **18**(2), 14. doi:10.5751/ES-05404-180214

Douglas J (2011) *Learning from Country: The Value of Indigenous Language and Culture Programs in Remote Schools for Community Engagement and Natural Resource Management*. Ninti One Ltd, Alice Springs. <http://www.nintione.com.au/resource/NintiOneResearch Report_69_Thevalueofcountryvisitsinremoteschools.pdf>.

Ferguson J (2012) A sustainable future for the Australian rangelands. *Rangeland Journal* **34**(1), 27–32. doi:10.1071/RJ11056

Fleming DA, Measham TG (2015) *Rich and Poor: Which Areas of Australia are Most Unequal?* The Conversation. 28 September 2015. <https://theconversation.com/rich-and-poor-which-areas-of-australia-are-most-unequal-42409>.

Gorecki S, Kelly J (2012) Treasury's wellbeing framework. *Economic Roundup* **2012**(3). <http://www.treasury.gov.au/PublicationsAndMedia/Publications/2012/Economic-Roundup-Issue-3/Report/Treasury8217s-Wellbeing-Framework>.

Harris S (1990) *Two-way Aboriginal Schooling: Education and Cultural Survival*. Indigenous Studies Press, Canberra.

Haynes RD (1998) *Seeking the Centre: The Australian Desert in Literature, Art and Film*. Cambridge University Press, Cambridge.

Healy MA (Ed.) (2015) *It's Hot and Getting Hotter: Australian Rangelands and Climate Change*. Reports of the Rangeland Cluster Project. Ninti One Ltd/CSIRO, Alice Springs.

Hill R, Pert PL, Davies J, Walsh F, Robinson C, Falco-Mammone F (2013) *Indigenous Land Management in Australia: Extent, Scope, Diversity, Barriers and Success Factors*. Report to Australian Landcare Council Secretariat, CSIRO Ecosystem Sciences, Cairns. <http://www.daff.gov.au/__data/assets/pdf_file/0010/2297116/ilm-report.pdf>.

Hinkson M, Smith B (2005) Introduction: conceptual moves towards an intercultural analysis. *Oceania* **75**(3), 157–166. doi:10.1002/j.1834-4461.2005.tb02877.x

Hoffmann BD, Roeger S, Wise P, Dermer J, Yunupingu B, Lacey D, Yunupingu D, Marika B, Marika M, Panton B (2012) Achieving highly successful multiple agency collaborations in a cross-cultural environment: experiences and lessons from Dhimurru Aboriginal Corporation and partners. *Ecological Management and Restoration* **13**(1), 42–50. doi:10.1111/j.1442-8903.2011.00630.x

Holcombe S, Yates P, Walsh F (2011) Reinforcing alternative economies: self-motivated work by central Anmatyerr people to sell Katyerr (desert raisin, bush tomato) in central Australia. *Rangeland Journal* **33**(3), 255–265. doi:10.1071/RJ10081

Holmes J (1997) Diversity and change in Australia's rangeland regions: translating resource values into regional benefits. *Rangeland Journal* **19**(1), 3–25. doi:10.1071/RJ9970003

Holmes J (2002) Diversity and change in Australia's rangelands: a post-productivist transition with a difference? *Transactions of the Institute of British Geographers* **27**(3), 362–384. doi:10.1111/1475-5661.00059

Holmes J (2010) The multifunctional transition in Australia's tropical savannas: the emergence of consumption, protection and Indigenous values. *Geographical Research* **48**(3), 265–280. doi:10.1111/j.1745-5871.2009.00629.x

Horstman M, Wightman G (2001) Karparti ecology: recognition of Aboriginal ecological knowledge and its application to management in north-western Australia. *Ecological Management and Restoration* **2**(2), 99–109. doi:10.1046/j.1442-8903.2001.00073.x

Jupp T, Fitzsimons J, Carr B, See P (2015) New partnerships for managing large desert landscapes: experiences from the Martu Living Deserts Project. *Rangeland Journal* **37**(6), 571–582.

Kondo N, Sembajwe G, Kawachi I, van Dam RM, Subramanian SV, Yamagata Z (2009) Income inequality, mortality, and self-rated health: meta-analysis of multilevel studies. *BMJ (Clinical Research Ed.)* **339**, b4471. doi:10.1136/bmj.b4471

Lea T (2008) *Bureaucrats and Bleeding Hearts: Indigenous Health in Northern Australia.* UNSW Press, Sydney.

Lee LS (2012) Horticultural development of bush food plants and rights of Indigenous people as traditional custodians – the Australian bush tomato (*Solanum centrale*) example: a review. *Rangeland Journal* **34**(4), 359–373. doi:10.1071/RJ12056

Maru Y, Davies J (2011) Supporting cross-cultural brokers is essential for employment among Aboriginal people in remote Australia. *Rangeland Journal* **33**(4), 327–338. doi:10.1071/RJ11022

Maru YT, Woodford K (2007) Revisiting sustainability boundaries from a systems perspective. *Modsim 2007: International Congress on Modelling and Simulation: Land, Water and Environmental Management: Integrated Systems for Sustainability*, pp. 477–482. <http://www.mssanz.org.au/MODSIM07/papers/8_s33/RevisitingSustainability_s33_Maru_.pdf>

Maru YT, Fletcher CS, Chewings VH (2012) A synthesis of current approaches to traps is useful but needs rethinking for indigenous disadvantage and poverty research. *Ecology and Society* **17**(2), 7. doi:10.5751/ES-04793-170207

McAllister RRJ, Smith DMS, Stokes CJ, Walsh FJ (2009) Patterns of accessing variable resources across time and space: desert plants, animals and people. *Journal of Arid Environments* **73**(3), 338–346. doi:10.1016/j.jaridenv.2008.10.007

Measham TG, Brake L, Robinson CJ, Larson S, Richards C, Smith TF (2011) NRM engagement between remote dryland communities and government agencies: success factors from Australia. *Journal of Arid Environments* **75**(10), 968–973. doi:10.1016/j.jaridenv.2011.04.018

Muller S (2012) 'Two ways': bringing Indigenous and non-Indigenous knowledges together. In *Country, Native Title and Ecology.* (Ed. J Weir) pp. 59–79. Australian National University e-pres/Aboriginal History Inc. (Monograph 24), Canberra.

Ngurruwutthun N (2014) My learning journey: two worlds leadership in Yirrkala. In *Northern Territory Principals: Educational Leaders Reflect on School Leadership.* (Ed. G Fry) pp. 13–17. Charles Darwin University, Darwin.

NTDET (Northern Territory Department of Education) (2011) Indigenous languages and culture. In *NT Curriculum Framework.* Northern Territory Department of Education, Darwin. <http://www.education.nt.gov.au/parents-community/curriculum-ntbos/ntcf>.

Ostrom E (2005) *Understanding Institutional Diversity.* Princeton University Press, Princeton.

Preuss K, Dixon M (2012) 'Looking after country two-ways': insights into Indigenous community-based conservation from the Southern Tanami. *Ecological Management and Restoration* **13**(1), 2–15. doi:10.1111/j.1442-8903.2011.00631.x

Prout S, Howitt R (2009) Frontier imaginings and subversive Indigenous spatialities. *Journal of Rural Studies* **25**(4), 396–403. doi:10.1016/j.jrurstud.2009.05.006

Rea N, Messner J (2008) Constructing Aboriginal NRM livelihoods: Anmatyerr employment in water management. *Rangeland Journal* **30**(1), 85–93. doi:10.1071/RJ07044

Resilience Alliance (2002) *Key Concepts*. Resilience Alliance, Stockholm. <http://www.resalliance.org/key-concepts>.

Robinson CJ, Wallington TJ (2012) Boundary work: engaging knowledge systems in co-management of feral animals on Indigenous lands. *Ecology and Society* **17**(2), 16. doi:10.5751/ES-04836-170216

Robinson C, Williams L, Lane MB (2009) A broker diagnostic for assessing local, regional and LEB-wide institutional arrangements for Aboriginal governance of desert environments. In *People, Communities and Economies of the Lake Eyre Basin*. (Eds T Measham, L Brake) pp. 217–249. Research Report 45. Desert Knowledge CRC, Alice Springs. <http://www.nintione.com.au/resource/DKCRC-Report-45-People-communities-and-economies-of-the-Lake-Eyre-Basin.pdf>.

Robinson CJ, Gerrard E, May T, Maclean K (2014) Australia's Indigenous carbon economy: a national snapshot. *Geographical Research* **52**(2), 123–132. doi:10.1111/1745-5871.12049

Rowley KG, O'Dea K, Anderson I, McDermott R, Saraswati K, Tilmouth R, Roberts I, Fitz J, Wang ZM, Jenkins A, Best JD, Wang Z, Brown A (2008) Lower than expected morbidity and mortality for an Australian Aboriginal population: 10-year follow-up in a decentralised community. *Medical Journal of Australia* **188**(5), 283–287.

Rudolph LI, Hoeber Rudolph S (1967) *The Modernity of Tradition: Political Development in India*. Chicago University Press, Chicago.

Russell-Smith J, Cook GD, Cooke PM, Edwards AC, Lendrum M, Meyer CP, Whitehead PJ (2013) Managing fire regimes in north Australian savannas: applying Aboriginal approaches to contemporary global problems. *Frontiers in Ecology and the Environment* **11**, e55–e63. doi:10.1890/120251

Salmon M, Gerritsen R (2013) A more effective means of delivering conservation management: a 'New Integrated Conservation' model for Australian rangelands. *Rangeland Journal* **35**(2), 225–230. doi:10.1071/RJ12080

Schwab RG, Sutherland D (2003) Indigenous learning communities: a vehicle for community empowerment and capacity development. *Learning Communities: International Journal of Learning in Social Contexts* **1**(1), 53–70.

Scott J (2012) *Social Network Analysis*. 3rd edn. Sage, London.

SCRGSP (Steering Committee for the Review of Government Service Provision) (2014) *Overcoming Indigenous Disadvantage: Key Indicators 2014*. Productivity Commission, Melbourne. <http://www.pc.gov.au/research/ongoing/overcoming-indigenous-disadvantage/key-indicators-2014#thereport>.

Sen A (1999) *Development as Freedom*. Oxford University Press, Oxford.

Sider GM (2014) Making and breaking the Aboriginal remote: realities, languages, tomorrows (a commentary). *Oceania* **84**(2), 158–168. doi:10.1002/ocea.5047

Smyth D (2011) *Review of Working on Country and Indigenous Protected Area Programs through Telephone Interviews*. Final report. Smyth and Bahrdt Consultants, Atherton. <https://www.environment.gov.au/indigenous/workingoncountry/publications/pubs/woc-interviews.pdf>.

Stafford Smith M (2008) The 'desert syndrome': causally linked factors that characterise outback Australia. *Rangeland Journal* **30**(1), 3–14. doi:10.1071/RJ07063

Stafford Smith M, Cribb J (2009) *Dry Times: Blueprint for a Red Land.* CSIRO Publishing, Melbourne.

Stafford Smith M, Moran M, Seemann K (2008) The 'viability' and resilience of communities and settlements in desert Australia. *Rangeland Journal* **30**(1), 123–135. doi:10.1071/RJ07048

Taylor J (2002) *The Spatial Context of Indigenous Service Delivery.* Working Paper 16/2002. Centre for Aboriginal Economic Policy Research, Australian National University, Canberra.

Taylor J (2009) Social engineering and Indigenous settlement: policy and demography in remote Australia. *Australian Aboriginal Studies* **1**, 4–16.

Taylor J (2011) Postcolonial transformation of the Australian Indigenous population. *Geographical Research* **49**(3), 286–300. doi:10.1111/j.1745-5871.2011.00698.x

Ten Deserts Initiative (2015) *Ten Deserts: Connecting People, Connecting Country.* Ten Deserts Initiative, Alice Springs. <http://tendeserts.org/>.

Tiplady T, Barclay MA (2007) *Indigenous Employment Evaluation (IEE) Tool.* Centre for Social Responsibility in Mining, University of Queensland, Brisbane. <https://www.csrm.uq.edu.au/docs/Evaluation%20Tool_WEB.pdf>.

Ursin H, Eriksen HR (2010) Cognitive activation theory of stress (CATS). *Neuroscience and Biobehavioral Reviews* **34**(6), 877–881. doi:10.1016/j.neubiorev.2009.03.001

Vinson T, Rawsthorne M (2015) *Dropping off the Edge 2015: Persistent Communal Disadvantage in Australia.* Jesuit Social Services/Catholic Social Services Australia, Melbourne.

Walker B, Salt D (2006) *Resilience Thinking: Sustaining Ecosystems and People in a Changing World.* Island Press, Washington DC.

Walker B, Holling CS, Carpenter SR, Kinzig A (2004) Resilience, adaptability and transformability in social-ecological systems. *Ecology and Society* **9**(2), 9. doi:10.5751/ES-00650-090205

Walker BW, Porter D, Marsh I (2012) *Fixing the Hole in Australia's Heartland: How Government Needs to Work in Remote Australia.* Desert Knowledge Australia, Alice Springs.

Walsh F, Davies J (2011) *Our Work is about Learning, Colleagues, Culture and Place: Aboriginal Employment at the Alice Springs Desert Park.* Ninti One Ltd, Alice Springs. <http://www.nintione.com.au/resource/NintiOneResearchReport_72_AboriginalemploymentatAliceSpringsDesertParkLR.pdf>.

Walsh F, Douglas J (2011) No bush foods without people: the essential human dimension to the sustainability of trade in native plant products from desert Australia. *Rangeland Journal* **33**(4), 395–416. doi:10.1071/RJ11028

Wenger E (1999) *Communities of Practice: Learning, Meaning, and Identity.* Cambridge University Press, Cambridge.

White N, Ober R, Frawley J, Bat M (2009) Intercultural leadership: strengthening leadership capabilities for Indigenous education. In *Australian School Leadership Today.* (Eds NC Cranston, LC Ehrich) pp. 85–103. Australian Academic Press, Brisbane.

Woinarski J, Traill B, Booth C (2014) *The Modern Outback: Nature, People and the Future of Remote Australia.* Pew Charitable Trusts. <http://www.pewtrusts.org/~/media/Assets/2014/10/TheModernOutbackForWeb.pdf>.

Woodward E (2008) Social networking for Aboriginal land management in remote northern Australia. *Australasian Journal of Environmental Management* **15**(4), 241–252.

Woolcock M, Narayan D (2000) Social capital: implications for development theory, research, and policy. *World Bank Research Observer* **15**(2), 225–249. doi:10.1093/wbro/15.2.225

Young E, Doohan K (1989) *Mobility for Survival: A Process Analysis of Aboriginal Population Movement in Central Australia.* North Australia Research Unit, Australian National University, Darwin.

Zander KK (2013) Understanding public support for Indigenous natural resource management in northern Australia. *Ecology and Society* **18**(1), 11. doi:10.5751/ES-05267-180111

6
Sustainability science, place and regional differences: vulnerability and adaptive capacity in Sydney

Tom Measham, Bruce Taylor and David Fleming

Geographers have long argued that 'to be … is to be *somewhere*' (Bondi and Davidson in Morrissey *et al*. 2014, p. 121). In this chapter we extend this principle a modest step to say that 'to be sustainable, is to be sustainable *somewhere*'. In so doing, the chapter demonstrates the crucial importance of place and region as fundamental concerns of sustainability science. The regional scale represents a key interface where biophysical, social and economic factors interact – it is the junction between local concerns and global influences. The chapter builds on the foundational work of sustainability science authors who developed frameworks to explore place-based vulnerability and adaptive capacity within a regional context. We present a case study of this framework looking at risk events in coastal Sydney, and consider how geographers and governance scholars have explored cross-scale linkages. Looking to other contexts where disasters have been studied in more detail, we further demonstrate that social relations within place-based communities are continually being renewed and that they can change substantially following disaster events. It is important to view place-based systems as dynamic and in constant interaction with their wider context.

Sustainability science, place and region

The concept of 'place' is intrinsically embedded in the sustainability science discourse. Consider, for example, one of the core questions posed by sustainability science:

> *What determines the vulnerability or resilience of the nature–society system* in particular kinds of places *and for particular types of ecosystems and human livelihoods?* (emphasis added, Kates et al. 2001, p. 641).

The local scale represents the meeting ground for interacting biophysical, social and economic forces which shape the material character and subjective human experience of particular places. Another of the foundation works in sustainability science is the framework developed by Turner *et al*. (2003). This framework was developed in response to the increasing need to address the consequences of global environmental changes as they manifest at local and regional scales (Adger and Kelly 1999). Of fundamental importance to this work was the recognition that vulnerability stems not only from exposure to hazards but also from the sensitivity and adaptability of local and regional systems, thereby bringing into focus a constellation of social, economic and geographic dimensions. This framework emphasises the geographic nature of vulnerability, with a nested set of global, regional and place-specific factors which influence vulnerability, as a set of interlinked 'boxes within boxes'. Moreover, the framework draws attention to two-way influences, such that changes in the social system at the local scale may trigger changes in social systems at higher scales, depending on how higher-level governments respond to local or regional events (Turner et al. 2003).

This classic framework for vulnerability analysis in sustainability science has been presented as a stylised diagram without reference to any particular case (Turner et al. 2003). In this chapter we have redrawn the framework to focus on a case study of vulnerability to extreme heat events in coastal Sydney, Australia (Fig. 6.1). This framework demonstrates that the geographical concept of place is, quite literally, at the centre of vulnerability analysis. It is important to recognise that places are not uniform in their scale and composition – a place represents a coherent subsystem bringing together dimensions of exposure, sensitivity and adaptive capacity. Exposure characteristics are likely to include elements of the physical environment such as watercourses, elevation above sea level and temperatures. Sensitivity relates to the social characteristics of the human population – questions related to human health, age and social capital. As demonstrated in Fig. 6.1, these human conditions interact with environmental conditions such that people with relative advantage or disadvantage may be more or less vulnerable due to their immediate local environmental conditions (and their past experiences of risk in that place or experiences of risk brought with them from other places). The elderly may be more sensitive to extreme temperatures;

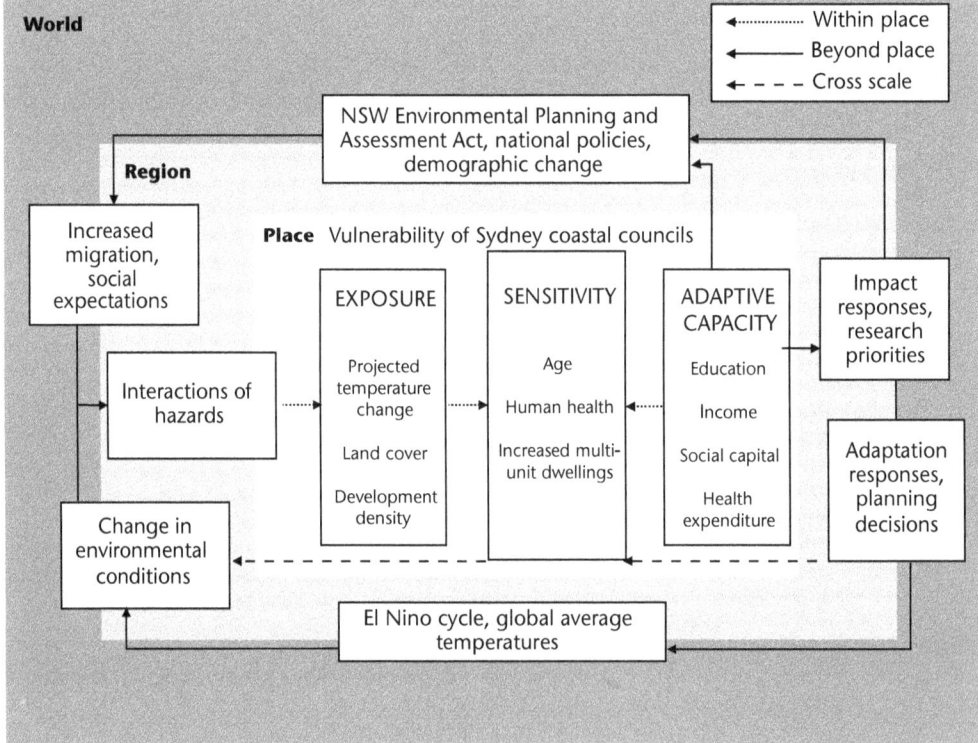

Fig. 6.1. Vulnerability to extreme heat in Sydney. Source: Adapted from original figure in Turner et al. (2003). ©(2003) National Academy of Sciences, USA.

however, this can be mediated by an ocean breeze or exacerbated by an urban heat island effect. It is worth noting at this point that the interactions between human conditions and environmental conditions is at the heart of the diagram, which highlights that questions of place, vulnerability and sustainability science in general are intrinsically concerned with the interaction.

The final part of the place subsystem on the right-hand side of the inner box is focused on adaptive capacity. As demonstrated by Fig. 6.1, adaptive capacity is fundamentally concerned with human responses. The figure shows three classes of responses, which all contribute to the adaptive capacity dimension of vulnerability. These include (i) coping response, i.e. the ability to weather a storm or survive a drought; (ii) impact response, which we may think of in terms of conventional emergency response; and (iii) adjustment/adaptation. The final class of response is arguably the most important and the most challenging aspect of vulnerability. It is concerned with longer-term rethinking and restructuring of the place subsystem, fundamentally addressing the relationship between human conditions and environmental conditions.

The concept of social vulnerability recognises sensitive populations that may be less able to respond to and recover following a natural disaster (Cutter and Finch 2008). In addition to the elderly, other groups that may be relatively more sensitive to the impacts of extreme events include the homeless, visitors and transient populations, disabled people and non-native speakers (Wisner 1998; Morrow 1999). Gender is also recognised as affecting vulnerability, with women being more vulnerable than men to disasters in a wide range of contexts (Rygel et al. 2006; Neumayer and Plümper 2007). This is partly due to disparities in income and education, with single mothers being particularly vulnerable (Kleinosky et al. 2007). In addition to increased vulnerability, women often experience the impacts of disasters disproportionately, as they are more likely to hold less secure jobs which can disappear after a disaster strikes (Rygel et al. 2006). However, it is also worth noting that in some cases women may be more resilient than men with regard to natural disaster preparation and recovery (Enarson and Morrow 1998). For these reasons, vulnerability research has recognised the increased vulnerability of marginalised groups and sought to map these factors as part of sustainability science (Preston et al. 2011).

Investigating vulnerability in coastal Sydney

A collaboration between researchers and local authorities initially took place from 2006 to 2008, with a follow-up study in 2011, to understand and respond to climate risks at the local to regional scales (Measham and Preston 2012; Taylor et al. 2013). The first phase of the project involved defining and mapping vulnerability to climate impacts across the 15 participating local governments. This was initially conducted as a desktop exercise and then presented to council representatives and staff for feedback, resulting in revisions to the assessment. Five areas of potential climate impacts were considered for the vulnerability assessment:

1. extreme heat;
2. bushfire (wildfire);
3. ecosystems and natural assets;
4. sea-level rise;
5. extreme rainfall and flooding.

Drawing on the framework depicted in Fig. 6.1, vulnerability was conceptualised as having three components: exposure, sensitivity and adaptive capacity (Preston et al. 2008).

It is important to clarify that what the assessment process calculated was relative vulnerability, involving a ranking process for how vulnerable each council was relative to the others. Indicators of exposure, sensitivity and adaptive capacity were integrated within a geographic information system to produce maps for each type of vulnerability by location, and a composite map which demonstrated net relative vulnerability across the study area. The research team compared their assumptions underpinning the calculations with those of Sydney Coastal Councils Group (SCCG) member council staff, who confirmed that they seemed reasonable. For some threats, such as sea-level rise and bushfires, vulnerability maps generally agreed well with the risk perceptions of council staff. However, council staff often didn't have sufficient knowledge of other threats to form an opinion on the appropriateness of the maps. This suggested that councils were far more aware of and sensitive to risks for which they had direct management authority or historical experience, or on which there were vocal community concerns. The results of this phase pointed to several hotspots within the Sydney region that were considered relatively more at-risk from the effects of climate change than other locations (Preston *et al.* 2008).

This phase resulted in three key findings. The first was to recognise the significant spatial variability across the SCCG region with respect to climate change vulnerability. The different classes of climate impacts varied from highly fragmented to concentrated in certain areas. This suggested the need to tailor responses to accommodate the unique challenges posed by different impacts across the area. The second was that demographic and socio-economic characteristics of the particular local populations within each of the council areas were as relevant as biophysical hazards when determining the potential for harm, reinforcing the importance of social vulnerability (Cutter and Finch 2008). The third was that the process of conducting the assessment was just as important as the outcome. Defining appropriate indicators of exposure, sensitivity and adaptive capacity from literature and comparing them with council officers' perspectives based on lived experience served as comparison and validation processes across different types of knowledge (Measham and Preston 2012).

In a second phase of the project, 15 workshops were conducted (one with each participating local government area), to consider the results of the vulnerability mapping process in relation to each council. The workshops included a participatory exercise to identify opportunities for and barriers to action. The workshops were attended by a broad cross-section of roles within local government including elected councillors representing local constituents, and council staff such as social planners, land-use planners, lawyers, engineers, senior managers, environmental officers and community engagement specialists (Smith *et al.* 2008). The workshops emphasised three themes: community capacity and expectations, infrastructure vulnerability, and barriers to and challenges for municipal planning.

To gain a deeper understanding of each of the cross-cutting issues that emerged from the workshops, a series of interviews was designed to consider each theme. Thirty-three semi-structured interviews with representatives from the councils were conducted in April and May 2008. A sample was designed to ensure the questions being considered covered a wide range of perspectives including those of elected councillors, senior managers, middle managers and operational staff. The interviews demonstrated that the participating councils had already made significant progress in addressing climate change through greenhouse gas mitigation efforts. For some time, councils had been engaged in efforts to reduce emissions from council facilities and community constituents. This reflected the participating councils' widespread awareness of climate change, and the growing momentum for substantive actions to reduce emissions. Such efforts did not then, however, address the issue of adapting to the effects of climate change that cannot be avoided through mitigation. Local governments' more recent efforts on climate change adaptation illustrate the evolution of thinking and policy around emergent issues of public concern. Such efforts were tentative and *ad hoc*, composed of a mixture of community engagement and geotechnical risk assessment. Interviews with council staff and councillors provided a clear indication that, generally, the participating councils would like to exercise greater leadership in ensuring communities are appropriately prepared.

Participants acknowledged responsibility for a range of climate-related impacts including the need to revise details of stormwater runoff and the approach to coastal inundation. Participants also noted major barriers to climate adaptation emanating from the state and federal policy environments in which local government operates. At the time of the interviews, New South Wales state legislation and management frameworks relevant to local government activities assumed a stable climate. As a consequence, there was little ability for councils to manage climate risks within the frameworks that they use. This was exacerbated by the fact that other legislation placed restrictions on local government authority and decision-making with respect to building codes, rate increases, and limits on growth and development. Collectively, these issues hampered progressive action by local government with respect to climate adaptation.

The interviews recognised that municipal planning represents a key avenue for local adaptation, but is subject to a wide set of recognised and unrecognised constraints. The former includes acknowledged issues such as limited resources and lack of information. By contrast, the under-acknowledged issues include local leadership, unconducive institutional context and competing planning agendas (Measham *et al.* 2011). On the one hand, councils had limited capacity to cope with a broad range of regulatory and service demands imposed through state legislation, given limited available resources including financial capital, technical information and expertise. On the other hand, local government operations were structured around thematic 'silos', which compartmentalised expertise in core operational areas, and limited internal deliberation and diffusion of knowledge. By working closely with the councils and local stakeholders, the research served to reinforce the understanding of these actors and to provide them with new information and ways of thinking about vulnerability and adaptation in order to reduce exposure and increase adaptive capacity.

Sustainability science highlights that challenges such as adaptation are not separable from the cultural, political, economic, environmental and developmental contexts in which they occur (Adger *et al.* 2013; Wise *et al.* 2014). In our case study, considering the project in relation to the framework set out in Fig. 6.1, the process of understanding the different dimensions of vulnerability at the local scale (inner box) was very useful for the project partners. It also gave them a scientific basis and justification for seeking change in the operating environment (Region box). Most notably, the project partners used the findings of the project not only in building innovation, but also in state and national policies. The most tangible example of success in this area was successfully lobbying the state government to alter the planning framework with relation to sea-level rise (Measham *et al.* 2011). This achievement reminds us of the importance of cross-scale linkages when considering vulnerability analysis and sustainability science more generally. Geographers have long focused on these linkages and drawn attention to the ways in which different actors approach linkages between local places and across scales in the context of sustainability science.

Linkages across places

Building local adaptive capacity through networks of extra-local cooperation is a strategy commonly used by local governments in climate mitigation and adaptation (Castán Broto and Bulkeley 2013; Fünfgeld 2015; Harman *et al.* 2015b). The practice of cooperation among neighbouring local governments in Australia, more broadly, is also well established. In recent decades this has taken the form of voluntary regional organisations of councils (Marshall *et al.* 2003; Dollery *et al.* 2012). These groupings were formed, initially with support of the Australian federal government, to help address extra-local planning, development and service provision problems that required cross-boundary coordination, and to encourage resource-sharing and economies of scale. They have also served, at times, as part of a deliberate political strategy by local actors to resist forced amalgamations and the imposition of inappropriate 'regional' boundaries or programs by higher-level governments (Herrschel 2007; Taylor 2012). In these cases, place and the maintenance of shared local identities contribute to endogenous regionalism, which can help or hinder institutionalisation of sustainability (Taylor 2012). Indeed, the contribution of the extra-local networks becomes clearer in their context of broader multilevel governance arrangements. In some regions of Australia, including our case of the Sydney coastal councils, the linkages are being employed in the governance of environmental risk, including attempts to reduce local vulnerability to climate hazards such as sea-level rise and inundation from storm surge.

In a follow-on study with the SCCG councils undertaken in 2011, researchers again adopted a participatory approach to examine the suitability of different planning responses to the specific risk of coastal inundation (Taylor *et al.* 2013). Several tensions relating to place and the relationship between local, regional and higher-level actors were observed. One tension related to the technical assessment of inundation risk within and between local council areas. Previous work had identified that inundation risk varied spatially across council areas due to differences in exposure, sensitivity and adaptive capacity. However, councils' methods to model, map and interpret that risk also varied. Council planners, engineers and councillors reported that the absence of a regionally consistent and agreed methodology increased the likelihood of conflict with developers and residents. They believed that different interests sought to play-off the findings of one risk assessment in one location with another in neighbouring locations, where similar development proposals were rated differently. Councils also discussed the implications of adopting different approaches to down-zoning 'at-risk' land or changing other property rights (e.g. setting time-based conditions for occupying land). Along with differences in risk assessment methods, these were seen to create an uneven, and unfair, distribution of costs and benefits between councils in the region, both economically and politically. In response, the councils argued for a strategy of 'scaling-up' the risk assessment process. This would require pooling financial and technical resources to cooperatively design and implement a regionally consistent method. Key to the successful adoption of the agreed method was enlisting state government endorsement of that assessment method. This, they argued, would ensure that local decisions were nested within a higher level of authority, reducing the risk of legal challenge (and financial liability) to local councils working in isolation. Further, the calls for consistency between councils were matched with arguments (by the same councils) for maintaining local discretion in the development approval process and in the use of local planning instruments. Central to these arguments was the objective of maintaining capacity to make place-based judgements or exercise discretion based on local knowledge and physical conditions.

The experiences of councils in the Sydney region echo the challenges of responding to sea-level rise faced by coastal local authorities in many parts of the globe. The success of action to reduce risk depends on the extent to which it can be tailored to suit the physical and institutional conditions of local places, the degree to which costs and benefits are seen to be fairly distributed between neighbouring areas, and the degree to which place-based responses can be embedded, and therefore legitimated, within higher levels of governmental authority (Fletcher *et al.* 2014; Harman *et al.* 2015a).

Exposure to hazards changes social relations

Social capital affects adaptive capacity and is an important component of preparing for and responding to natural disasters (Adger *et al.* 2005). However, it is important to recognise that social capital is complex (Pelling and High 2005) and can change in the context of natural disasters. Although it has been established that people affected by natural disasters demonstrate increased trust towards strangers (Andrabi and Das 2010; Cassar *et al.* 2011), more recently an additional complication has been observed: trust within place-based communities can be negatively affected by natural disasters even while trust towards strangers may increase (Fleming *et al.* 2014).

A range of factors is thought to differently affect the behaviour of people in the short and long term following disasters. Cassar *et al.* (2011) describe different conditions that can positively affect people's levels of trust after natural disasters: more community affinity given the necessity to work together to recover from damages, more solidarity between neighbours, perceptions that new natural disasters may occur and therefore the support of others in the future becomes important, and a decrease in the degree of income disparity (so income inequality decreases). On the other hand, Fleming *et al.* (2014) described aftermath conditions from the 2011 Chilean earthquake that negatively affected the levels of trust within communities and provided important insights for potential effects on adaptive capacity elsewhere. Specifically, the Chilean example revealed the importance of rivalry generated by disputes to obtain scarce relief and recovery recourses, migration (movement between communities), social displacement of people (movement within communities), and increased information asymmetries between neighbours (a type of aftermath moral hazard). A key lesson for vulnerability in Sydney and elsewhere is that asymmetric information flows following natural disasters, regarding damages suffered and income losses across neighbours, can strain previously established social relations (Fleming *et al.* 2014).

Implications for sustainability science

When considering the broader spatial context of vulnerability in Fig. 6.1 (middle and outer zones), there is a wide range of considerations, both human and non-human, all in continuous change, reflecting both human power relationships and natural forces (Massey 1994). Within these spaces, particular places represented by the inner box are local manifestations of the wider forces in the middle and outer zones. Due to the scale differences between places and spaces, each particular place may not express all of the social relations that have influenced it. This is particularly relevant to cases of local expression of global aspects of sustainability science (Leitch and Robinson 2012). Though a town, suburb or rural property is unique, its local identity is intrinsically tied to a broader context affected by issues such as global markets, concerns for biodiversity, sea-level rise and social movements (Abel et al. 2011; Measham and Lockie 2012). As demonstrated through our Sydney case study, focusing on the linkages across places reveals how local actors shape new institutions and learn from each other across space (Measham and Preston 2012; Taylor et al. 2013).

Considering the effects of natural hazards on social relations is important to better understand how the social and biophysical dimensions of place systems co-evolve. The example from Chile shows how multiple social effects can occur. Relating these to Fig. 6.1, it is clear that two distinct types of impact responses were noted following the disaster: there was a decline in reciprocity among villagers in affected areas, but an increase in trust towards strangers (Fleming et al. 2014). This shows the complex ways in which local experiences of hazards are entwined within broader social relations.

Conclusion

This chapter started from the premise that to be sustainable is to be sustainable somewhere, just as 'to be unsustainable is to be unsustainable somewhere'. Places – be they coastal towns or culturally significant landscapes – represent systems whose core is the relationship between humans and environmental conditions. This approach to thinking about sustainability science demonstrates a clear link between social systems, culture and biophysical systems. Moreover, the chapter has demonstrated a strong relationship between exposure to risk and social capital, such that vulnerability is intrinsically tied to social relations which are continuously evolving and entwined with biophysical factors. The chapter has further drawn attention to linkages across places and across scales, such that local places are not simply subsystems of global forces. Rather they are interconnected and have the potential to learn from each other, reducing their own vulnerability and potentially triggering change at the state or national scale.

Acknowledgements

Thanks to Carol Farbotko, Ben Harman and Karin Hosking for helpful comments on an earlier version of this chapter.

References

Abel N, Gorddard R, Harman B, Leitch A, Langridge J, Ryan A, Heyenga S (2011) Sea level rise, coastal development and planned retreat: analytical framework, governance principles and an Australian case study. *Environmental Science and Policy* **14**(3), 279–288. doi:10.1016/j.envsci.2010.12.002

Adger WN, Kelly PM (1999) Social vulnerability to climate change and the architecture of entitlements. *Mitigation and Adaptation Strategies for Global Change* **4**(3–4), 253–266. doi:10.1023/A:1009601904210

Adger WN, Hughes TP, Folke C, Carpenter SR, Rockström J (2005) Social-ecological resilience to coastal disasters. *Science* **309**(5737), 1036–1039. doi:10.1126/science.1112122

Adger WN, Barnett J, Brown K, Marshall N, O'Brien K (2013) Cultural dimensions of climate change impacts and adaptation. *Nature Climate Change* **3**(2), 112–117. doi:10.1038/nclimate1666

Andrabi T, Das J (2010) In Aid We Trust: Hearts and Minds and the Pakistan Earthquake of 2005. Policy Research Working Paper No. WPS 5440. World Bank, Washington DC.

Cassar A, Healy A, von Kessler C (2011) Trust, risk, and time preferences after natural disasters: experimental evidence from Thailand. Unpublished manuscript.

Castán Broto V, Bulkeley H (2013) A survey of climate change experiments in 100 cities. *Global Environmental Change* **23**, 92–102. doi:10.1016/j.gloenvcha.2012.07.005

Cutter SL, Finch C (2008) Temporal and spatial changes in social vulnerability to natural hazards. *Proceedings of the National Academy of Sciences of the United States of America* **105**(7), 2301–2306. doi:10.1073/pnas.0710375105

Dollery B, Grant B, Kortt M (2012) *Councils in Cooperation: Shared Services and Australian Local Government.* Federation Press, Sydney.

Enarson E, Morrow BH (1998) *The Gendered Terrain of Disasters.* Praeger, Westport, CT.

Fleming DA, Chong A, Bejarano HD (2014) Trust and reciprocity in the aftermath of natural disasters. *Journal of Development Studies* **50**(11), 1482–1493. doi:10.1080/00220388.2014.936395

Fletcher CS, Taylor BM, Rambaldi AN, Harman BP, Heyenga S, Ganegodage KR, Lipkin F, McAllister RR (2014) Equity, economic efficiency and institutional capacity in adapting coastal settlements. In *Applied Studies in Climate Adaptation.* (Eds JP Palutikof, SL Boulter, J Barnett, D Rissik) pp. 208–215. John Wiley & Sons, New York.

Fünfgeld H (2015) Facilitating local climate change adaptation through transnational municipal networks. *Current Opinion in Environmental Sustainability* **12**, 67–73. doi:10.1016/j.cosust.2014.10.011

Harman BP, Heyenga S, Taylor BM, Fletcher CS (2015a) Global lessons for adapting coastal communities to protect against storm surge inundation. *Journal of Coastal Research* **31**(4), 790–801. doi:10.2112/JCOASTRES-D-13-00095.1

Harman BP, Taylor BM, Lane MB (2015b) Urban partnerships and climate adaptation: challenges and opportunities. *Current Opinion in Environmental Sustainability* **12**, 74–79. doi:10.1016/j.cosust.2014.11.001

Herrschel T (2007) Regions between imposed structure and internally developed response: experiences with twin track regionalisation in post-socialist eastern Germany. *Geoforum* **38**, 469–484. doi:10.1016/j.geoforum.2006.11.003

Kates RW, Clark WC, Corell R, Hall JMI, Jaeger CC, Lowe I, McCarthy JJ, Schellnhuber HJ, Bolin B, Dickson NM, Faucheux S, Gallopin GC, Grübler A, Huntley B, Jäger J, Jodha NS, Kasperson RE, Mabogunje A, Matson P, Mooney H, Moore B, O'Riordan T, Svedin U (2001) Sustainability science. *Science* **292**(5517), 641–642. doi:10.1126/science.1059386

Kleinosky LR, Yarnal B, Fishe A (2007) Vulnerability of Hampton Roads, Virginia to storm-surge flooding and sea-level rise. *Natural Hazards* **40**, 43–70. doi:10.1007/s11069-006-0004-z

Leitch AM, Robinson CJ (2012) Shifting sands: uncertainty and a local community response to sea level rise policy in Australia. In *Risk and Social Theory in Environmental Management.* (Eds T Measham, S Lockie) pp. 117–131. CSIRO Publishing, Melbourne.

Marshall N, Dollery B, Witherby A (2003) Regional Organisations of Councils (ROCs): the emergence of network governance in metropolitan and rural Australia? *Australasian Journal of Regional Studies* **9**, 169–188.

Massey D (1994) *Space, Place and Gender.* Polity Press, Cambridge.

Measham T, Lockie S (Eds) (2012) *Risk and Social Theory in Environmental Management.* CSIRO Publishing, Melbourne.

Measham T, Preston BL (2012) Vulnerability analysis, risk and deliberation: the Sydney Climate Change Adaptation Initiative. In *Risk and Social Theory in Environmental Management.* (Eds T Measham, S Lockie) pp. 147–157. CSIRO Publishing, Melbourne.

Measham TG, Preston BL, Smith TF, Brooke C, Gorddard R, Withycombe G, Morrison C (2011) Adapting to climate change through local municipal planning: barriers and challenges. *Mitigation and Adaptation Strategies for Global Change* **16**, 889–909. doi:10.1007/s11027-011-9301-2

Morrissey J, Nally D, Strohmayer U, Whelan Y (2014) *Key Concepts in Historical Geography*. Sage, London.

Morrow BH (1999) Identifying and mapping community vulnerability. *Disasters* **23**(1), 1–18. doi:10.1111/1467-7717.00102

Neumayer E, Plümper T (2007) The gendered nature of natural disasters: the impact of catastrophic events on the gender gap in life expectancy, 1981–2002. *Annals of the Association of American Geographers* **97**(3), 551–566. doi:10.1111/j.1467-8306.2007.00563.x

Pelling M, High C (2005) Understanding adaptation: what can social capital offer assessments of adaptive capacity? *Global Environmental Change* **15**(4), 308–319. doi:10.1016/j.gloenvcha.2005.02.001

Preston BL, Smith T, Brooke C, Gorddard R, Measham TG, Withycombe G, McInnes K, Abbs D, Beveridge B, Morrison C (2008) *Mapping Climate Change Vulnerability in the Sydney Coastal Councils Group*. Prepared for the Sydney Coastal Councils Group and the Australian Greenhouse Office, Melbourne.

Preston BL, Yuen EJ, Westaway RM (2011) Putting vulnerability to climate change on the map: a review of approaches, benefits, and risks. *Sustainability Science* **6**(2), 177–202. doi:10.1007/s11625-011-0129-1

Rygel L, O'Sullivan D, Yarnal B (2006) A method for constructing a social vulnerability index: an application to hurricane storm surges in a developed country. *Mitigation and Adaptation Strategies for Global Change* **11**(3), 741–764. doi:10.1007/s11027-006-0265-6

Smith TF, Preston B, Gorddard R, Brooke C, Measham TG, Withycombe G, Beveridge B, Morrison C (2008) *Regional Workshops Synthesis Report: Sydney Coastal Councils' Vulnerability to Climate Change. Part 1*. Report prepared for the Sydney Coastal Councils Group. CSIRO Climate Adaptation Flagship, Melbourne.

Taylor BM (2012) Regionalism as resistance: governance and identity in Western Australia's wheatbelt. *Geoforum* **43**, 507–517. doi:10.1016/j.geoforum.2011.11.004

Taylor BM, Harman BP, Inman M (2013) Scaling-up, scaling-down, and scaling-out: local planning strategies for sea-level rise in New South Wales, Australia. *Geographical Research* **51**(3), 292–303. doi:10.1111/1745-5871.12011

Turner BL, Kasperson RE, Matson PA, McCarthy JJ, Corell RW, Christensen L, Eckley N, Kasperson JX, Luers A, Martello ML, Polsky C, Pulsipher A, Schiller A (2003) A framework for vulnerability analysis in sustainability science. *Proceedings of the National Academy of Sciences of the United States of America* **100**(14), 8074–8079. doi:10.1073/pnas.1231335100

Wise RM, Fazey I, Stafford Smith M, Park SE, Eakin HC, Archer van Garderen ERM, Campbell B (2014) Reconceptualising adaptation to climate change as part of pathways of change and response. *Global Environmental Change* **28**, 325–336. doi:10.1016/j.gloenvcha.2013.12.002

Wisner B (1998) Marginality and vulnerability: why the homeless of Tokyo don't 'count' in disaster preparations. *Applied Geography (Sevenoaks, England)* **18**(1), 25–33. doi:10.1016/S0143-6228(97)00043-X

7

A hierarchy of needs for achieving impact in international Research for Development

James R.A. Butler, Toni Darbas, Jane Addison, Erin L. Bohensky, Lucy Carter, Michaela Cosijn, Yiheyis T. Maru, Samantha Stone-Jovicich, Liana J. Williams and Luis C. Rodriguez

In developing countries, change in environmental and socio-economic systems is occurring at unprecedented rates, driven by rapid globalisation, technological advances, modernisation and increasingly unpredictable economic and environmental shocks (Leach 2008). As a consequence, there are growing concerns that conventional international development initiatives aiming to reduce poverty, conserve or sustainably use the environment and increase resilience, are becoming less effective. Recently, Ramalingam (2013) argued that aid programs' established assumptions of linear, simple cause-and-effect relationships which have long guided interventions and their evaluation are no longer valid. Instead, he argues for a more 'systemic, adaptive, networked, dynamic approach … and a fundamental shift in the mental models, strategic approaches, organisational philosophies and performance approaches of foreign aid' (Ramalingam 2013, p. 361).

This has important implications for the framing and practice of research that aims to support international development. In scientific terms, analysis of many development problems requires a complex adaptive systems approach, which considers that most of the key challenges we face today (e.g. food insecurity, poverty, climate change, deforestation) are embedded in nonlinear and path-dependent systems that are characterised by multiple components, drivers of change emanating at different scales (e.g. global, regional, local) and sectors (e.g. agriculture, health and infrastructure), multiple equilibrium points, feedback loops and thresholds where abrupt and irreversible change can occur (Lansing 2003; Scoones 2009).

Following Snowden and Boone (2007), we consider that development problems can be characterised as complex, complicated or chaotic (Fig. 7.1). Complicated problems such as managing multispecies fisheries assume linear direct and indirect cause-and-effect relationships (e.g. more captures result in more income) and nonlinear behaviour (e.g. stock dynamics) which require multidisciplinary approaches. Complex problems such as food insecurity or climate adaptation involve stakeholders from multiple levels and sectors, and hence require transdisciplinary science that engages their diverse knowledge types to understand the systems and to accommodate their varied perceptions of causes and solutions. Chaotic issues, such as war or natural disasters, have an immediate requirement for action, and sci-

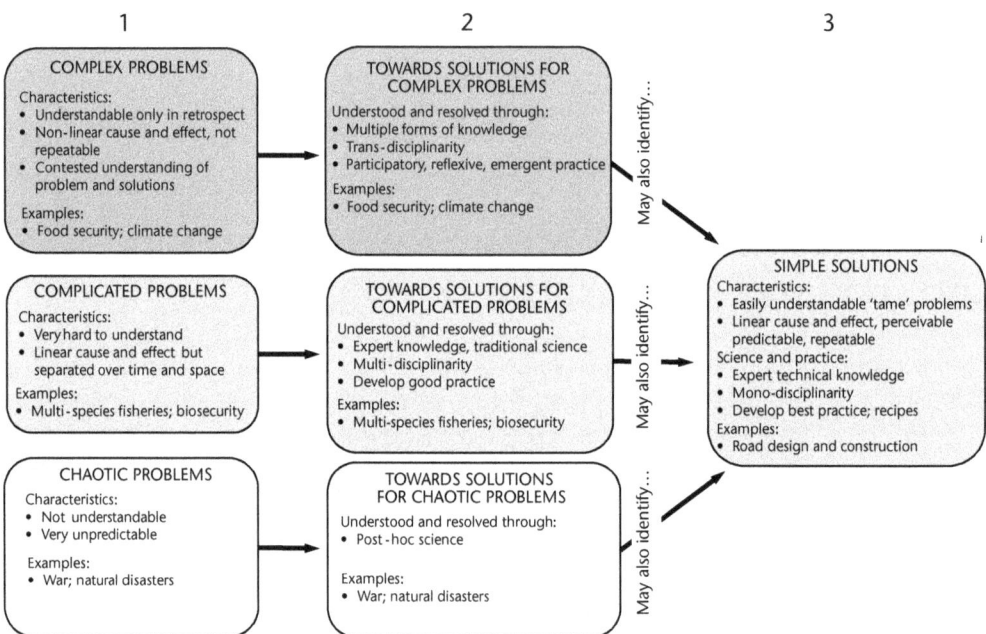

Fig. 7.1. A typology of framings for (1) international development problems, (2) appropriate science practice and (3) possible simple approaches and practice which may be identified by (2).

ence is likely to be *post hoc*. All these processes may identify some components of problems as being simple; these can then be tackled through conventional monodisciplinary approaches applying technical knowledge and best practice, reflecting more typical research.

Addressing complex development problems requires scientists to adopt alternative practices. This change in emphasis has been termed 'research for development' (R4D), whereby research pro-actively seeks to bring about positive change by brokering different knowledge types, of which science is only one, and building capacity of all actors in the system through learning and innovation processes (Hall *et al.* 2003; Hawkins *et al.* 2009). Most importantly, scientists must regard themselves as active components of the system concerned, rather than as passive observers (Ballard 2005; Cundill and Fabricius 2010; Butler *et al.* 2016a), who aim to generate transformative or systemic change which achieves scale and sustainability of development outcomes beyond the temporal and spatial limits of a project (Fowler and Dunn 2014).

As a research organisation with a remit to inform government policy including aid, CSIRO has the skills base and opportunity to excel in R4D. However, only modest steps have been made in this regard (Stone-Jovicich *et al.* 2015). This chapter describes the authors' approach to the science and practice of R4D which redefines CSIRO's contribution to Australia's Official Development Assistance (ODA) program. We propose a 'hierarchy of needs' for R4D to have impact on complex development problems. Using examples from our team's recent projects, we highlight lessons learned from applying our R4D approach, and key research questions that have emerged.

A hierarchy of needs approach in practice

Our hierarchy is presented in Fig. 7.2. The team's ultimate goal is to support Australia's ODA objective, which reflects global development targets such as the Millennium

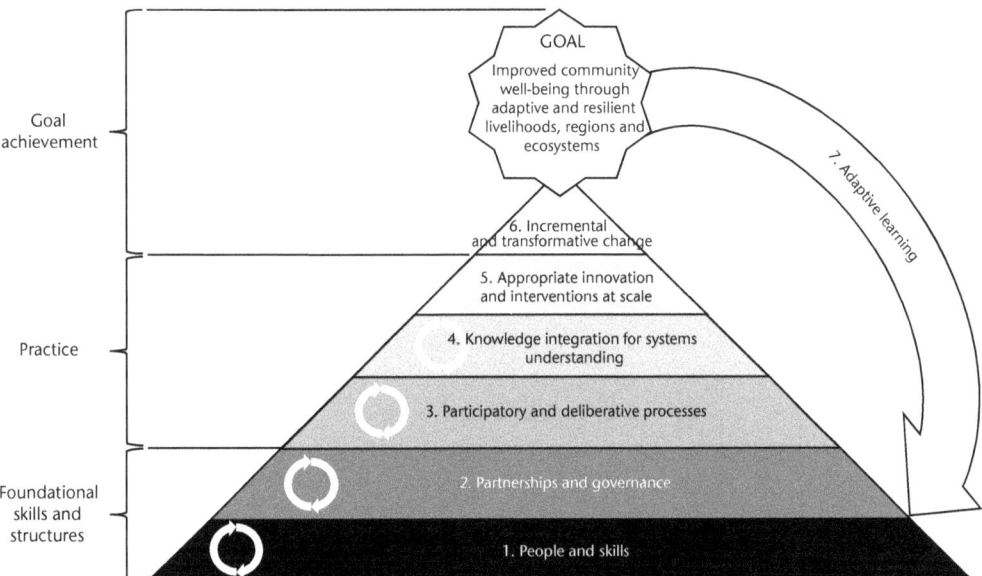

Fig. 7.2. The levels of the R4D process framed as a hierarchy of needs. Circular arrows within each level indicate ongoing adaptive learning by the team and stakeholders.

Development Goals (UN 2014) and their successors, the Sustainable Development Goals (UN 2015), summarised as 'improved community well-being through adaptive and resilient livelihoods, regions and ecosystems'. To achieve this, six components or levels must be considered: (1) people and skills, (2) partnerships and governance, (3) participatory and deliberative processes, (4) systems understanding, (5) appropriate innovation and interventions at scale, and (6) incremental and transformative (or systemic) change. Levels 1 and 2 are the foundational skills and structures which enable the sequential practice of Levels 3, 4 and 5. If those are met then it is possible to achieve the goal, which includes Level 6. A seventh level – adaptive learning – creates a feedback loop through monitoring and evaluation which can test the hierarchy's assumptions and redesign activities within each level.

We discuss each level in sequence, applying the following steps. First, we present a proposition that is 'an attempt to make explicit our present understanding and our underlying mental models of change' (Walker et al. 2006, p. 1). Then, we provide brief observations from seven recent R4D projects tackling complex problems (Table 7.1). Finally, we present key knowledge gaps and research questions that should be addressed to refine the approach. We accept that these levels are not always discrete, and the tight linkages between them are evident in the overlaps in our case study examples. Nonetheless, we have found this to be a useful way of organising a description of our team's approach.

People and skills (Level 1)
Proposition: 'Research teams require multiple science disciplines plus brokering, facilitation, communications and conflict management skills.'

For R4D to effectively address complex problems, there must be reconsideration of the way that researchers and their organisations view and undertake science. Methods and tools which put ideas and knowledge into productive use, incorporate facilitation methods and tools, and adopt a willingness to experiment with new ways of working are more likely to

Table 7.1. Summaries of the seven case study projects discussed

Case study project and acronym	Location	Goal	Funders and partners	Timeframe
Food systems innovation (FSI) initiative	Australia and a selection of ODA-supported countries	Improve the impact of ODA-supported food and agricultural programs in the Indo-Pacific by enabling innovation in Australian and international food systems development and practice	AusAID/DFAT; AusAID/DFAT; ACIAR, CSIRO and in-country DFAT and ACIAR programs and local partners	2012–15
Applied research and innovation systems in agriculture (ARISA)	Indonesia	Establish partnerships between the Indonesian research sector and private sector to generate innovation in food production and distribution	DFAT; Indonesian universities and private sector	2014–17
Climate futures and rural livelihood adaptation strategies in Nusa Tenggara Barat Province (NTB), Indonesia (NTB Climate Futures)	Indonesia	Increase the adaptive capacity of rural communities, boundary partners and researchers to enable vulnerability reduction to adverse change	AusAID and DFAT; universities, national and provincial government research agencies, NGOs, private sector	2010–14
Climate futures, ecosystem services and rural livelihood adaptation strategies in West New Britain, Papua New Guinea (WNB Climate Futures)	Papua New Guinea	Increase the adaptive capacity of rural communities, government and other stakeholders to enable vulnerability reduction to adverse change	DFAT and Department of the Environment; national and provincial government departments, NGOs, private sector	2012–13
Sustainable and resilient farming system intensification (SRFSI)	India, Bangladesh, Nepal	Enhance the resilience of rice-based farming communities in the East Gangetic Plains by intensifying crop production	ACIAR; agricultural research institutes, universities, government extension services, NGOs	2014–17
Developing multiscale adaptation strategies for farming communities (ACCA)	Cambodia, Laos, Bangladesh, India	Improved farming livelihoods, increased household resilience and more flexibility to respond to climate variability in rice-based farming systems	ACIAR; agricultural research institutes, universities, government extension services, NGOs	2010–15
Strengthening incentives for improved grassland management in China and Mongolia (SIIGM)	Mongolia, China	National-level policy recommendations on incentives and subsidies for improving pastoral livelihoods and landscape condition	ACIAR; agricultural research institutes, universities, local ministries	2014–19

ACIAR = Australian Centre for International Agricultural Research; DFAT = Department of Foreign Affairs and Trade.

bring about systemic change (Jerneck *et al.* 2011). This necessitates the involvement of individuals from a range of science disciplines, plus capacities that are not conventionally incorporated into research teams (Stone-Jovicich *et al.* 2015). In particular, brokering for improved knowledge and partnerships is fast becoming recognised as an essential skill for addressing the complexities of food insecurity (Klerkx and Gildemacher 2012) and climate adaptation (Butler *et al.* 2016a).

Examples

The Food Systems Innovation (FSI) initiative aimed to enhance the design and implementation of Australian food security-related ODA in the Indo-Pacific region by fostering innovation among Australian and international partners through the exchange of knowledge, learning and networking (Table 7.1). Running for 3.5 years from 2012 to 2015, it involved three Australian government agencies with roles in food and agricultural systems research, policy and practice: the AusAID/Department of Foreign Affairs and Trade (DFAT), the Australian Centre for International Agricultural Research (ACIAR) and CSIRO. By taking a complex systems approach to institutional change and innovation, FSI attempted to bridge various epistemological and practice boundaries in the three-partner organisations. In order to foster innovation, the CSIRO team (composed primarily of multidisciplinary scientists) had to shift from being exclusively researchers to also being brokers of concepts, knowledge, and networks.

Brokering was challenging and resource-intensive because many of the concepts and practices introduced in FSI were novel to partners. One of these was nutrition-sensitive agriculture (NSA), which attempts to incorporate nutritional trade-offs and considerations into agricultural development to address the growing economic, social and health problems caused by malnutrition (Gómez *et al.* 2013). The FSI team had to not only facilitate the involvement of international expertise that was regarded as legitimate by all partners, but also to communicate NSA cross-culturally within the organisations and their overseas offices and collaborators. Additionally, the team had to encourage participation and learning, which was time-consuming and taxing. While FSI achieved success in brokering acceptance and adoption of NSA tools and approaches among the partners, this came at the expense of the researchers producing traditional science outputs.

A second example is a three-country initiative aiming to provide national-level policy recommendations for improving pastoral livelihoods and landscape condition on the Mongolian plateau (SIIGM; Table 7.1). The formal inclusion of a scientist-cum-broker has already contributed to increased cross-disciplinary awareness and cross-cultural understanding between country teams about their varied agrarian histories and path-dependencies. While the broker's role will diminish with time, in the interim the responsibility for managing any failed relationships falls largely on this person. Also, the broker's key role in identifying and responding to shifting power and political dynamics through time can detract from their formal scientific role.

Response to the proposition

Brokering for knowledge integration and transformative partnerships (see Level 2) and closely related skills such as cross-sectoral communication, mediation, conflict resolution, motivational coaching, event organisation and facilitation are key to catalysing innovation to bring about change in complex systems. We consistently found that particular personality styles and skills (e.g. adaptability, conflict aversion) contribute positively to brokering. These are not necessarily held by all researchers, and cannot be easily transferred through training. Brokering as a researcher in a science organisation requires a specific set of individual and organisational conditions before it can be successful. A cultural shift in how science values and supports these skills is necessary.

Research questions
Three questions emerge from our experiences.

1. How can science organisations better support researchers who broker knowledge, resolve conflicts and facilitate to bring about change?
2. What influence does diversity (of age, gender and ethnicity), individually and in teams, have on scientists who undertake brokering roles?
3. How can the capacity of researchers to become better knowledge-brokers and facilitators be improved, particularly within developing countries where capacity may be lower and cultural norms may differ?

Partnerships and governance (Level 2)
Proposition: 'Meaningful partnerships must be mutually beneficial and must be built based on negotiations and agreements for their implementation and maintenance.'

Complex problems inherently involve multiple interests and stakeholders (Fig. 7.1). Recently, there has been increasing attention to the development of intersectoral partnerships between stakeholders from the private sector, government and civil society (Bitzer 2012). To catalyse innovation, partnerships are central to institutional arrangements necessary for the generation of knowledge and building capacity to engender systemic change (Glasbergen 2007). However, partnering is complex in itself, with each stakeholder bringing personal and organisational agendas to the table. Therefore, it is critical that early in R4D projects meaningful and mutually beneficial partnerships are negotiated and agreed between stakeholders including researchers, then maintained through learning and adaptive governance that enables adjustment to the changing needs of the partnership (Hall 2011) and the development of an appropriate exit strategy when the partnership is no longer needed.

Examples
Despite a long history of interactions and common projects between the three organisations involved in FSI, a partnership agreement was not established until mid-2014 to help partners define a common vision of project goals and approaches, and to better formalise organisational roles and clarify expectations. The partnership was reviewed regularly to build trust and to enable FSI to adapt to shifts in ODA policy, overseen by a steering committee which provided strategic direction, and a management committee which was responsible for day-to-day operations. All partners were represented on both committees.

A second example, the Applied Research and Innovation Systems in Agriculture (ARISA) project, was initiated by DFAT and CSIRO in Indonesia in 2014 (Table 7.1). Through a competitive grant process, it establishes partnerships between Indonesian research institutions and the private sector to commercialise existing research and create a two-way dialogue to generate innovation in food production and distribution. Capacity-building of all grant partners on how to establish and maintain collaboration through the establishment of partnership agreements, dialogue and reflective processes is a core component of ARISA. This process is co-facilitated and brokered by an Indonesian consultant and a CSIRO research scientist specialising in partnership processes. Each partnership is establishing its own governance structure tailored to the intervention, innovation process and number of partners. As a minimum, they are establishing biannual strategic meetings with all partners and monthly meetings for day-to-day decision-making.

Response to the proposition

While partnerships are a crucial starting point in R4D projects, their ability to tackle complex problems varies according to their characteristics. There are a variety of partnership types on a continuum from transactional to collaborative and transformative (Butcher et al. 2011; Gray and Stites 2013). Transactional types tend to be partnerships of convenience, driven by individual organisational purposes. However, they are often limited in terms of institutional change as they tend not to address power imbalances (Horton et al. 2009). Collaborative partnerships focus on close working relationships, combining diverse skills and co-creating joint action, which results in long-term commitment. Transformational partnerships generate innovation to modify the system through policy dialogue, and institutional and organisational change which influences a project's sustainability (Butcher et al. 2011; Austin and Seitanidi 2012; Gray and Stites 2013).

Partnerships are not static but evolve, and to effectively address complex problems transactional and collaborative types must become transformative. This requires governance which enables ongoing renegotiation of the partnership and the management of power relations, ensuring accountability and transparency, plus negotiation of mutual benefits (Horton et al. 2009). This is not trivial in terms of skills and time. Impartial brokers (as noted in Level 1) are crucial to build the organisational relations within the partnership, ensure commitment of resources, and solve problems.

Research questions

Although multi-stakeholder partnerships have become a core component of dealing with complex problems, there continues to be confusion about their definition, value and efficacy. There are still gaps in understanding how partnerships function. There is a body of work on what constitutes good practice (e.g. Horton et al. 2009; Dodds 2015; Hazlewood 2015) but it is often difficult to translate into action. In addition, no two partnerships or their governance structures are the same – they are highly context-specific. This raises three questions.

1. How can partnerships be evaluated to assess their functioning and sustainability?
2. How can the capacity of stakeholders be built to engage in functional partnerships that mature and adapt through learning and reflection processes?
3. What are the governance principles required to enable transformative partnerships to emerge?

Participatory and deliberative processes (Level 3)

Proposition: 'Stakeholder participation in research is imperative to promote legitimacy, ownership, double- and triple-loop learning and action.'

Participation encompasses a range of meanings and engagement processes, from the provision of information to handing over power to stakeholders (Leitch et al. 2015). Participation can encourage transparency, share diverse types of knowledge, create empowerment, build trust, legitimise decision-making, and allow social learning to occur through deliberation among participants (Brown 2008; Reed et al. 2010). To address the systemic root causes of community vulnerability to shocks and stressors, which are frequently institutional and political (Pelling 2011; Rodima-Taylor et al. 2012), participatory processes must be designed to identify and challenge these issues through double-loop learning (revisiting assumptions about cause and effect) and triple-loop learning (reassessing underlying values and beliefs, potentially resulting in changes to institutional norms; Pahl-Wostl 2009). Because participatory research is complex and often occurs over multiple years, it is important to track outcomes and impacts systematically (Cundill and Fabricius 2009; Fazey et al. 2014; Oteros-Rozas et al. 2015, Butler et al. 2016a, b). However, participatory processes are also fraught with power dynamics which can undermine the intended outcomes (Parfitt 2004; Hurlbert and Gupta 2015; Butler et al. 2015). Rather than empower marginalised actors, they may further subjugate them through a 'tyranny of participation' (Cooke and Kothari 2001).

Examples

Between 2010 and 2014, we explicitly aimed to evaluate multi-stakeholder adaptation planning processes in two projects which examined community development trajectories and adaptation pathways (Table 7.1): NTB Climate Futures (Indonesia) and WNB Climate Futures (Papua New Guinea). The projects were founded on participatory principles which aimed to provide forums for double- and triple-loop learning and knowledge co-generation through a series of multi-stakeholder workshops which we co-facilitated with research partners (Butler et al. 2015). We also designed an evaluation framework to measure the projects' outcomes and impacts. This was based on the principles of adaptive co-management (Cundill and Fabricius 2009, 2010), whereby the participatory process was assumed to enhance stakeholders' capacity for systemic change by generating innovation and institutional change through enhanced leadership, trust, new social networks and partnerships, social learning and systems understanding, power-sharing and conflict resolution (see Level 7; Butler et al. 2016a).

Some key lessons emerged. First, the ease of engaging stakeholders varies according to the context. In WNB, all the invited community-level participants and NGOs attended, often accompanied by other relations and friends. However, the attendance of government and private-sector stakeholders was poor. In NTB, where it is standard practice to offer *per diem* payment to participants, community attendance was also very high and, unlike WNB, government and NGO attendance was almost 100%, which generated more extensive social networks among participants. As a consequence, the dynamics of the learning processes and their outcomes varied. It was impossible to objectively investigate the reasons for invitees' attendance rates, but as noted elsewhere (e.g. Stringer et al. 2006; Armitage et al. 2008), they were potentially influenced by different incentives for participation, including competing work or family commitments, 'consultation fatigue' and perceived dis-benefits.

Second, in spite of our roles as facilitators and brokers engaged 'within' the system, understanding and negotiating politics and power in cross-cultural contexts was problematic. Our evaluation methods could not fully deconstruct the subtleties of power and influence both among participants and between participants and researchers, and their influence on outcomes (Butler et al. 2016a, b, c). Third, the type and longevity of participation is important. To achieve outcomes and impact, stakeholder engagement has to be maintained beyond the brief set of workshops and beyond the project's completion. Our evaluations showed that the workshops generated some multi-loop learning and enabled change agents willing to champion action, but this did not necessarily lead to immediate actions and outcomes (Butler et al. 2016a, b, c; Wise et al. 2016).

Response to the proposition
Significant investments of time, effort and resources are required to enable effective participation. As an example, a pertinent observation is the '10 to 1 rule': it takes up to 10 days to organise one day of workshops, including issues such as logistics, stakeholder analyses, human research ethics approvals and local permissions. Also, in spite of extensive planning, there is uncertainty about who ultimately attends and the political agendas that they bring, which are often influenced by unexpected events that affect workshop attendance. Consequently, an overreliance on workshops can lead to participation becoming a hostage to fortune. On the other hand, workshops are often the only practical way to catalyse and support collective learning and foster relationship-building. However, modifying established participatory methods to be more logistically or culturally appropriate can also create trade-offs with methodological robustness. These issues are compounded in trans-disciplinary, -cultural and -language contexts, or where participatory processes are novel. Finally, while participatory processes such as workshops may catalyse the emergence of change agents, learning and new social networks, participation has to extend beyond such one-off events and even the project lifetime (Butler *et al.* 2015, a, c, Stone-Jovicich *et al.* 2015). This requires alternative approaches, for example mentoring of change agents, embedding participation in longer-term statutory processes (e.g. government development planning), and establishing knowledge networks and appropriate media channels.

Research questions
Various priority issues arise from this brief analysis.

1. How can conventional models of stakeholder engagement be improved to promote legitimacy, ownership, double- and triple-loop learning and action?
2. If models of participation are limited to individual rather than group engagement (e.g. workshops), is social learning lost or compromised?
3. How do perceptions and forms of participation vary between cultural contexts, and which modes of engagement and learning are most effective in different contexts?
4. How can evaluation methods identify and quantify the influence of power, politics and incentives on participatory processes and their outcomes?

Knowledge integration for systems understanding (Level 4)
Proposition: 'Systems understanding based on multiple knowledge types is necessary to design appropriate interventions.'
A key prerequisite to understanding a system of interest is its definition and delineation (Walker *et al.* 2012). This refers to identifying the key components (biophysical, economic, social and political) of a system, how they interact, and what future that pattern of interaction is likely to deliver (Ramalingam 2013). The conceptual approaches to defining and diagnosing a system vary. For food systems they can include social-ecology, political ecology, innovation systems and agro-ecosystem framings individually or in combination (Foran *et al.* 2014). Regardless of which approach is applied, designing an intervention to shift a system towards a more desirable path requires the identification of its components and primary drivers, variables and feedbacks that determine its structure and function.

This requires engagement with multiple knowledge types, the nature of which is twofold. First, it refers to scientific knowledge garnered by deploying relevant disciplinary specialities which must be integrated in a mutually informative, transdisciplinary manner.

Second, it refers to garnering relevant non-scientific contextual knowledge (e.g. from extension officers, local government officials, resource users) that must be integrated both with each other and with scientific knowledge (Stock and Burton 2011). Integration of knowledge types through deliberative processes (Gauri et al. 2013) also enables social learning (see Level 3). This is particularly effective when stakeholders bring their differing perceptions of cross-scale system drivers and the 'option space' for responding: local and community-level actors are more likely to understand immediate and context-specific issues, while national government or NGO actors have a better understanding of longer-term, national and global-level drivers (Scoones 2009; Butler et al. 2014; Bohensky et al. 2016).

Examples

The Sustainable and Resilient Farming System Intensification (SRFSI) project focuses on the food systems in the Eastern Gangetic Plains that span north-east India, north-west Bangladesh and the eastern plains of Nepal (Table 7.1), one of the most food-insecure regions of the world (Erenstein and Thorpe 2011; Sugden 2009, 2013). The project aims to increase cropping intensity from subsistence levels to profitable second and third irrigated dry season crops. Although the same body of agronomic knowledge is relevant across the region, each nation has its own agrarian history, policies and extension programs that condition its delivery to farming communities which are themselves diversified by ethnicity, language and religion (Banerjee et al. 2014). SRFSI's design has taken a systems approach by analysing the dynamics of agrarian change across the region to identify pivotal bottlenecks and entry points to agricultural intensification, such as the feminisation of agriculture due to male labour migration (Lahiri-Dutt 2014), and applying innovation systems thinking (Hall et al. 2003) to harness and encourage institutional change. Innovation platforms which build interactions among agricultural system actors (Boogaard et al. 2013) have been convened to transform farming practices and governance by coordinating and leveraging multilevel stakeholders (e.g. input, service and credit providers, and public, private and civic sector extension services). Deep contextual knowledge is assured by in-country teams drawn from NGOs, national agricultural research councils, universities and departments and their various extension programs. Transnational disciplinary knowledge (e.g. in crop modelling, rural sociology, water management, food policy and wheat and maize improvement) is provided by CSIRO, Australian universities and international research institutes.

Experiences from SRFSI have highlighted that, as noted by Sumberg et al. (2013), the politics of system definition, disciplinary traditions and hierarchies and organisational funding can initially form barriers to integrating knowledge to develop a common understanding of the system, and consequently the ability of the project to deliver systemic change. This was evident during a prolonged debate regarding SRFSI's scaling-up strategy (see Level 5), considered unsatisfactory by the funders (ACIAR). Reformulation of the innovation platforms to include a broader range of stakeholders was necessary to resolve the issue. The necessity and value of CSIRO researchers' appropriate brokering and facilitation skills, effective partnerships and governance and participatory, multi-stakeholder processes was emphasised during this redesign (see Levels 1, 2 and 3).

Response to the proposition

In the case of the East Gangetic Plains, adopting a systems understanding has been crucial due to the multiscale complexity of the geopolitical, cultural and agrarian history overlaying the ecological system. SRFSI began fieldwork in mid-2014, but it was only towards the end of 2015 that the multiple partners and their knowledge types began to coalesce into a transdisciplinary diagnosis of the key barriers to intensification. The preparation of a scaling proposal contributed to a deliberative process to reflect on the value of innovation platforms, which facilitated the development of a systems view of the problem. In the end, the scaling proposal was not funded due to DFAT funding cuts. However, this did not undermine the development of trust and an effective partnership between the diverse stakeholders.

Research questions

Our experience in SRFSI highlights three critical questions.

1. How can epistemological and institutional barriers to building system understanding be overcome more effectively?
2. How can the politics of system definition be better navigated to achieve inclusive and equitable R4D?
3. How can novel forms of knowledge extension emerging through information technology be integrated through transdisciplinary R4D processes?

Appropriate level of intervention and innovation (Level 5)

Proposition: 'For research projects to support development impact, interventions should be implemented at the right and appropriate level.'

An appropriate intervention can be defined as one that balances the need to fit within socio-cultural systems or aspirations, and prompts some aspect of beneficial systemic change even beyond the life of the project. This expectation relates to ethical research practice (a demonstration of 'giving back' to communities which support and contribute to research) and accountability (responsible use of public funds; Cramb 2000).

Various strategies are used to plan for expanding the impact of interventions. In conventional linear approaches to agricultural development research, there has been a focus on supporting widespread dissemination and adoption of 'proven' agricultural, fisheries or aquaculture technologies, such as through provincial extension programs or farmer field days – referred to as 'scaling-out' (Millar and Connell 2010). This replicates technologies or processes demonstrated to have positive impact in a particular site, to other similar sites. In recognition of some of the limitations to this approach, there has been increased attention on the institutional changes required to foster and support local-level change over wide geographic contexts, such as informing policy mechanisms and creating links across value chains – referred to as 'scaling-up' (Millar and Connell 2010). This more systems-based approach, as outlined in Level 4 above, focuses more on sharing or pooling locally specific insights and lessons to inform a much wider policy and institutional context for change (Hall et al. 2010). The latter also relates to tackling the deeper underlying political and institutional drivers of community vulnerability (Pelling 2011; Rodima-Taylor et al. 2012), and can be termed transformative change (see Level 6).

Examples

The Adapting to Climate Change in Asia (ACCA) project sought to improve household livelihoods by building adaptive capacity to climate variability in rain-fed rice-producing areas of Laos, Cambodia, Bangladesh and India (Table 7.1). The project aimed to directly benefit the communities it was working with, and use outputs and lessons to inform higher-level policy and extension programs. Local-level on-farm research occurred in parallel with ongoing stakeholder engagement, including multiple government departments and NGOs. The intention was to build relationships and share learnings and insights from the project to identify opportunities for broader application, or higher-level constraints (e.g. access to input markets) that could be ameliorated through policy mechanisms or partnering with other programs. In this way, ACCA attempted to follow two impact pathways which tackled different levels of the food systems in each country: the farm level and the policy level (Roth and Grünbühel 2012; Stone-Jovicich et al. 2015).

At the farm level, adaptation practices and approaches tested through the project have been incorporated into local extension and NGO programs, resulting in the adoption and scaling-out of labour-saving devices and appropriate technology. At the policy level, ongoing efforts to inform broader adaptation policies and programs have been varied, largely due to rapidly changing contexts and the political nature of policy and program development. For example, the bifurcation of the Indian state of Andhra Pradesh resulted in significant uncertainty and creation of a new government administration; in Laos, promising avenues for supporting policy and research dialogue were cut under changes to the Australian ODA program.

Response to the proposition

Processes for taking innovation and impact to scale, either through dissemination or through facilitating systemic change, vary and need to evolve as situations change. Taking research to scale can be seen simplistically, tied to assumptions that, if benefits are demonstrated, wider application will follow. At the local level, this can be confounded, for example because farmer adoption can be limited by access to inputs or markets. In ACCA, embedding research outputs into NGO programs was a key strategy to ensure that access to information and inputs endured and were available in new areas.

Though achieving impact at scale is important, there needs to be constant adaptive learning and reflection by project governance to ensure the appropriateness of broader application of interventions, or strategies to manage for unanticipated consequences (see Level 2). In ACCA this was achieved through regular reflections with in-country project teams, participating farmers and policy and NGO partners. The adaptation practices themselves ranged from building on local knowledge, to changes to crop patterns and mechanisation. A household typology was developed to understand which households would be able to adopt new practices, and to consider which households might be negatively affected by adaptation practices (e.g. landless rural labour; Williams et al. 2015).

Research questions

The experience from ACCA raises three key questions.

1. How can short-term project cycles be extended to maintain sustained stakeholder engagement which can enable long-term policy and institutional change?
2. How do the benefits and impacts of innovations change with greater scales of application and adoption?
3. Who are the winners and losers from interventions at different scales, and what mechanisms can be put in place to detect and rectify unintended consequences?

Incremental and transformative change (Level 6)
Proposition: 'Complex development problems require incremental plus transformative change.'

Complex development issues such as chronic poverty, food and nutrition insecurity, entrenched inequality, natural resource depletion and degradation persist in the face of rapid socio-economic, political, environmental and climatic change. The circular nature of behavioural and structural causes of complex problems is such that they often generate traps – self-reinforcing feedback loops that keep social-ecological systems in persistent undesirable states (Carpenter and Brock 2008), such as poverty (Carter and Barrett 2006). Traps can arise as a result of limits in capital, institutions and networks (Dasgupta 2007; Maru et al. 2012), and are maintained by feedbacks among underlying systemic variables or drivers (Walker et al. 2012).

Interventions to complex problems are broadly categorised as incremental or transformative. Incremental strategies work within the incumbent system or processes in the short term (Park et al. 2012), identifying ways that livelihood capitals can be best used to cope (Reed et al. 2013) and extending actions that already reduce losses to natural variations in climate (Kates et al. 2012). Examples include diversifying or intensifying current agricultural production, enhancing skills and education, or improving infrastructure. Transformative strategies tackle underlying causes of community vulnerability and traps, such as institutions, policies and power which determine their rights and participation (Pelling 2011). They 'promote fundamentally alternative forms of development from those described for each site as dominant' (Pelling 2011, p. 140), creating a new system or process (Park et al. 2012) and addressing the institutions that constrain and shape social behaviour and the rules that affect negotiation and the performance of power (Reed et al. 2013). Examples are new production systems, and modified or new formal and informal institutions. Transformative change can occur abruptly but it usually emerges over years or decades (Feola 2015). The levers or interventions that can effect transformation are context-specific, but learning, social innovation and novel institutional arrangements are common principles (Snowden and Boone 2007; Hall 2011; Leach et al. 2012).

Examples

The NTB Climate Futures project aimed to increase the adaptive capacity of rural communities, boundary partners and researchers to enable vulnerability reduction to adverse change (Table 7.1). A three-stage process of planning workshops was designed. Stage 1 analysed potential future development pathways for rural communities from the perspective of provincial-level stakeholders, and identified priority 'no regrets' strategies that would form adaptation pathways. Stage 2 replicated this process at the subdistrict and village level. In Stage 3, participants' perspectives and intervention priorities were integrated (Wise et al. 2016). The processes used a range of deliberative methods to promote social learning (Butler et al. 2015). In addition, to identify and tackle systemic issues potentially responsible for poverty traps, double- and triple-loop learning was encouraged (see Level 3) through scenario analysis and modelling of potential future social and environmental change on ecosystem services (Skewes et al. 2016) and adaptive capacity (Butler et al. 2016c).

Following the Stage 1 and 2 workshops, we categorised the drivers identified at the different levels as either proximate (e.g. droughts, farmers' skills, water availability) or systemic (e.g. corruption, lack of women's empowerment, land rights). We also categorised the 'no regrets' strategies as incremental or transformative, and repeated this for the Stage 3 workshops. Only 30% and 42% of the drivers from the Stage 1 and 2 workshops, respectively, were systemic. Of the Stage 1 strategies, 67% were incremental and 33% were transformative. In Stage 2, 84% were incremental and only 16% were transformative (Butler et al. 2016c). Following Stage 3, the proportions remained similar (Wise et al. 2016).

Response to the proposition

The results of the NTB Climate Futures project show that in spite of the identification of systemic drivers by both higher-level and local-level stakeholders, the majority of strategies were incremental and reflected standard development needs. A potential explanation is that the process was too brief to fully examine and acknowledge systemic issues and traps. The Stage 1 and 2 workshops provided only one iteration of the social learning cycle, followed by another in Stage 3. In a developing country context, participatory scenario planning represents only the first stage of a long-term capacity-building process which must be maintained to enable decision-makers to reflect and learn iteratively (Chaudhury et al. 2013; Vervoort et al. 2014) (this is discussed further in the following proposition). Nonetheless, the multilevel planning approach represented a transformative social innovation in itself, which empowered community members and provided a forum for the exchange of knowledge and ideas (Butler et al. 2015). If sustained, such grassroots platforms can catalyse and implement transformation (Leach et al. 2012). Incremental actions are essential to build the capacity and opportunities that will allow transformative interventions to subsequently emerge (Kates et al. 2012; Park et al. 2012).

Research questions
To fully understand the relative importance of incremental and transformative change, and how to achieve these in order to attain the R4D goal, three questions are pertinent.

1. How can R4D processes maintain and nurture the capacity to design and implement transformational interventions after a project's completion?
2. How can R4D projects prepare stakeholders to take opportunities for transformation during 'windows of opportunity' such as shocks to the system?
3. How much incremental change is required to catalyse transformative change?

Adaptive learning (Level 7)
Proposition: 'Stakeholders, including researchers, must adopt appropriate monitoring, evaluation and learning to measure and continually refine the R4D process.'

Complex development problems are by their nature unpredictable and involve multiple scales, actors, sectors, drivers and feedback loops. Furthermore, no two problems are alike (Fig. 7.1). Consequently, it is imperative that, to achieve its goal, R4D must constantly question and respond to change. This relates both to the individual levels in the hierarchy, and to a project's influence on outcomes, impacts and the goal (Fig. 7.2), much of which will occur after the life of a project (see Level 5). Hence monitoring, evaluation and learning (MEL) tools must be designed to support project management, but also to understand and respond to emergent properties as the system evolves with the R4D process. If undertaken as a participatory process among researchers and other partners and stakeholders, and embedded within a project's governance structure (see Level 2), MEL can enhance and catalyse social learning, systems understanding, trust and institutional change. Sometimes referred to as a component of adaptive co-management, MEL can enable transformative change while also building the capacity and preconditions for long-term adaptive governance (Armitage et al. 2008; Cundill and Fabricius 2009, 2010; Butler et al. 2016a, b).

Examples
Designing MEL approaches and tools to guide and facilitate adaptive governance and management was a key component of the FSI initiative. Because of its multi-stakeholder and organisational complexity, incorporating methods to monitor and assess progress and learn along the way was crucial if FSI was to maintain relevance and achieve impact.

The team developed an MEL system to support the partners to achieve the initiative's objectives, while meeting the accountability needs of the funder (DFAT) and the committees (see Level 2) plus contributing to the different information, learning and management requirements of each partner and adapting to the shifting policy arena. One strategy was to include a theme on the concept of Theories of Change and Impact Pathways. These involve innovative tools to assist project managers and stakeholders plan and evaluate project progress, while testing assumptions and generating learning (Douthwaite *et al.* 2009; Vogel 2012), enabling the tools to be applied among FSI's overseas partners. Another was to constantly revise and adapt a suite of MEL tools to be 'fit-for-purpose' rather than 'best practice', and to embed them in the governance processes of the FSI committees.

Response to the proposition

Increasingly, experiences in implementing R4D projects highlight that the majority of existing MEL approaches and tools, in standard form or as stand-alone tools, are often unsuitable for the multiple accountability, learning and adaptive management needs and rhythms of the complex problems, partnerships and engagement processes which R4D projects are embedded in and shaped by. Many of the existing approaches in the literature and toolkits are excellent. However, most have been designed to either monitor tangible outputs or evaluate performance (primarily through linear approaches and indicator-based, quantitative methods), or to capture less tangible (less quantifiable) processes and dynamics of change (primarily narrative-based approaches). They tend to be robust for monitoring and evaluation purposes, but less suited for catalysing learning and adaptive co-management. While there is a growing body of work on learning-focused practices (e.g. Cundill and Fabricius 2009), these tend to be time- and data-intensive, expensive and highly context-dependent, and their value is difficult to communicate to funders.

Thus, while incorporating MEL into R4D is critical, identifying a suite of approaches and tools that are fit-for-purpose is a significant challenge. In putting together a set of MEL tools, the following principles need to be considered. They must:

- be useful for accountability purposes;
- foster reflection and learning among the R4D team members and partners;
- guide and enable adaptive co-management that can respond to short-term, immediate and emergent needs and opportunities;
- assist in the communication of key achievements, challenges and lessons learned, to different partners;
- capture both tangible and quantifiable dimensions (e.g. number of workshops) and less tangible and process dimensions (e.g. trust, leadership, partnerships);
- be flexible enough to respond to changes (e.g. policy shifts, unexpected problems).

Research questions

While there is a growing interest in MEL tools and processes to support R4D to achieve systemic and sustained impacts, the breadth and scope of research on the application of MEL remains relatively narrow. We highlight three research questions.

1. How, why and under what conditions are different types of MEL approaches and tools appropriate for and effective in R4D?
2. What types of MEL are best suited for different types of issues or problems being tackled by R4D projects?
3. How can MEL support the evaluation and enhancement of all levels in our hierarchy of needs?

Conclusion: understanding and solving complex development problems requires diverse research skills

As the complexity and range of international development problems increase, so the framing and practice of research supporting Australia's ODA program, and development programs globally, must evolve to deliver impact. While traditional, more 'bounded' forms of R4D which address simple and complicated problems remain relevant (Stone-Jovicich et al. 2015), supporting R4D which can understand and address complex problems is becoming a greater priority. Unlike bounded R4D, which is based on more structured, monodisciplinary and linear forms of technological innovation and transfer, R4D targeting complex problems is process-based. It places greater emphasis on achieving systemic and transformative change through social innovations such as novel learning and institutional arrangements.

Based on recent project experiences, which tackled complex food insecurity and climate adaptation problems in the Indo-Pacific region, we posit a hierarchy of needs to achieve impact in complex systems R4D. Our hierarchy suggests that the platform for the process consists of two basic levels: people and skills, and partnerships and governance. Only if these foundational skills and structures are established in an R4D project can the practice and goal achievement levels be attained. The seventh level of adaptive learning infuses each level with a reflexive feedback loop, which is applied across the R4D process as a whole. Rather than providing generic design principles (Coe et al. 2014; Stone-Jovicich et al. 2015), our hierarchy simply attempts to present the cumulative levels necessary to achieve the R4D goal in complex problems, based on our empirical experiences. The hierarchy itself offers a proposition to be tested and refined by other R4D scientists and practitioners.

Our brief exemplification and analysis of each level highlights several issues. First, there are clearly multiple overlaps and linkages between levels, and it is arguable how distinct some levels are (e.g. Level 3 participatory and deliberative processes, and Level 4 knowledge integration for systems understanding). Second, the examples reveal the common threads of people and skills (Level 1) and partnerships and governance (Level 2), in all levels. The research questions derived from our analysis of these deserve prioritisation if the capacity for R4D in CSIRO and similar organisations is to be enhanced. Third, the role of adaptive learning and governance, and the use of fit-for-purpose MEL tools and methods, are fundamental at each level and for the overall R4D process, to detect outcomes and impact. Again, research organisations should consider how to embed these skills and enhance this science capacity now there is growing interest in evaluating and quantifying research impact (Stone-Jovicich et al. 2015).

There is an important role for social and integrative scientists in leading and facilitating 'systems' or 'complexity' science for R4D. By applying their skills in social processes, governance and evaluation, these scientists are ideally placed to use participatory methods to integrate the multiple world views of other science disciplines and relevant stakeholders. However, many science organisations and funding bodies need to place greater value on these skills, which do not necessarily enable us to deliver impact in terms of technological fixes for simple problems, but position us to catalyse systemic change which may emerge only over lagged timeframes. Finally, it is worth considering that our R4D hierarchy of needs may be equally applicable to complex problems in the domestic arena.

Acknowledgements

We acknowledge funding support for the seven case studies from AusAID, DFAT, ACIAR and the Australian Department of the Environment. We also thank the myriad of Australian and international project partners who have silently contributed to this chapter and our hierarchy of needs.

References

Armitage D, Marschke M, Plummer R (2008) Adaptive co-management and the paradox of learning. *Global Environmental Change* **18**, 86–98. doi:10.1016/j.gloenvcha.2007.07.002

Austin JE, Seitanidi MM (2012) Collaborative value creation: a review of partnering between nonprofits and businesses. Part I. Value creation spectrum and collaboration stages. *Nonprofit and Voluntary Sector Quarterly* **41**, 726–758. doi:10.1177/0899764012450777

Ballard D (2005) Using learning processes to promote change for sustainable development. *Action Research* **3**, 135–156. doi:10.1177/1476750305052138

Banerjee O, Darbas T, Brown PR, Roth CH (2014) Historical divergence in public management of foodgrain systems in India and Bangladesh: opportunities to enhance food security. *Global Food Security* **3**, 159–166. doi:10.1016/j.gfs.2014.06.002

Bitzer VA (2012) Partnering for change in chains: the capacity of partnerships to promote sustainable change in global agrifood chains. *International Food and Agribusiness Management Review* **15**(Special Issue B).

Bohensky EL, Kirono D, Butler JRA, Rochester W, Habibi P, Handayani T, Yanuartati Y (2016) Climate knowledge cultures: stakeholder perspectives on change and adaptation in Nusa Tenggara Barat, Indonesia. *Climate Risk Management* **12**, 17–31. doi:10.1016/j.crm.2015.11.004

Boogaard B, Schut M, Klerkx L, Leeuwis C, Duncan A, Cullen B (2013) *Critical Issues for Reflection when Designing and Implementing Research for Development in Innovation Platforms*. Report prepared for the CGIAR Research Program on Integrated Systems for the Humid Tropics. Technology and Innovation Group, Wageningen University.

Brown VA (2008) *Leonardo's Vision: A Guide to Collective Thinking and Action*. Sense Publishers, Rotterdam.

Butcher J, Bezzina M, Moran W (2011) Transformational partnerships: a new agenda for higher education. *Innovative Higher Education* **36**, 29–40. doi:10.1007/s10755-010-9155-7

Butler JRA, Suadnya W, Puspadi K, Sutaryono Y, Wise RM, Skewes TD, Kirono D, Bohensky EL, Handayani T, Habibi P, Kisman M, Suharto I, Hanartani, Supartarningsih S, Ripaldi A, Fachry A, Yanuartati Y, Abbas G, Duggan K, Ash A (2014) Framing the application of adaptation pathways for rural livelihoods and global change in Eastern Indonesian islands. *Global Environmental Change* **28**, 368–382. doi:10.1016/j.gloenvcha.2013.12.004

Butler JRA, Wise RM, Skewes TD, Bohensky EL, Peterson N, Suadnya W, Yanuartati Y, Handayani T, Habibi P, Puspadi K, Bou N, Vaghelo D, Rochester W (2015) Integrating top-down and bottom-up adaptation planning: a structured learning approach. *Coastal Management* **43**, 346–364. doi:10.1080/08920753.2015.1046802

Butler JRA, Suadnya IW, Yanuartati Y, Meharg S, Wise RM, Sutaryono Y, Duggan K (2016a) Priming adaptation pathways through adaptive co-management: design and evaluation for developing countries. *Climate Risk Management* **12**, 1–16. doi:10.1016/j.crm.2016.01.001

Butler JRA, Bohensky EL, Darbas T, Kirono D, Wise RM, Sutaryono Y (2016b) Building capacity for adaptation pathways in eastern Indonesian islands: synthesis and lessons learned. *Climate Risk Management* **12**, A1–A10. doi:10.1016/j.crm.2016.05.002

Butler JRA, Bohensky EL, Suadnya W, Yanuartati Y, Handayani T, Habibi P, Puspadi K, Skewes TD, Wise RM, Suharto I, Park SE, Sutaryono Y (2016c) Scenario planning to leap-frog the Sustainable Development Goals: an adaptation pathways approach. *Climate Risk Management* **12**, 83–99. doi:10.1016/j.crm.2015.11.003

Carpenter SR, Brock WA (2008) Adaptive capacity and traps. *Ecology and Society* **13**(2), 40. http://www.ecologyandsociety.org/vol13/iss2/art40/

Carter MR, Barrett CB (2006) The economics of poverty traps and persistent poverty: an asset-based approach. *Journal of Development Studies* **42**(2), 178–199. doi:10.1080/00220380500405261

Chaudhury M, Vervoort J, Kristjanson P, Ericksen P, Ainslie A (2013) Participatory scenarios as a tool to link science and policy on food security under climate change in East Africa. *Regional Environmental Change* **13**, 389–398. doi:10.1007/s10113-012-0350-1

Coe R, Sinclair F, Barrios E (2014) Scaling up agroforestry requires research 'in' rather than 'for' development. *Current Opinion in Environmental Sustainability* **6**, 73–77. doi:10.1016/j.cosust.2013.10.013

Cooke B, Kothari U (2001) *Participation: The New Tyranny*. Zed Press, London.

Cramb R (2000) Evaluation soil conservation technologies: a summing up. In *Soil Conservation Technologies for Smallholder Farming Systems in the Philippine Uplands: A Socio-Economic Evaluation*. (Ed. RA Cramb) pp. 195–212. ACIAR Monograph No. 78. Australian Centre for International Agricultural Research, Canberra.

Cundill G, Fabricius C (2009) Monitoring in adaptive co-management: toward a learning based approach. *Journal of Environmental Management* **90**, 3205–3211. doi:10.1016/j.jenvman.2009.05.012

Cundill G, Fabricius C (2010) Monitoring the governance dimension of natural resource co-management. *Ecology and Society* **15**(1), 15. http://www.ecologyandsociety.org/vol15/iss1/art15/

Dasgupta P (2007) Nature and the economy. *Journal of Applied Ecology* **44**(3), 475–487. doi:10.1111/j.1365-2664.2007.01316.x

Dodds F (2015) *Multi-stakeholder Partnerships: Making Them Work for the Post-2015 Development Agenda*. Global Research Institute, University of North Carolina.

Douthwaite B, Alvarez S, Keatinge JDH, Mackay R, Thiele G, Watts J (2009) Participatory impact pathways analysis (PIPA) and research priority assessment. In *Prioritizing Agricultural Research for Development: Experiences and Lessons*. (Eds DA Raitzer, GW Norton) pp. 8–24. CABI, Wallingford.

Erenstein O, Thorpe W (2011) Livelihoods and agro-ecological gradients: a meso-level analysis in the Indo-Gangetic Plains, India. *Agricultural Systems* **104**, 42–53. doi:10.1016/j.agsy.2010.09.004

Fazey I, Bunse L, Msika J, Pinke M, Preedy K, Evely AC, Lambert E, Hastings E, Morris S, Reed MS (2014) Evaluating knowledge exchange in interdisciplinary and multi-stakeholder research. *Global Environmental Change* **25**, 204–220. doi:10.1016/j.gloenvcha.2013.12.012

Feola G (2015) Societal transformation in response to global environmental change: a review of emerging concepts. *Ambio* **44**, 376–390. doi:10.1007/s13280-014-0582-z

Foran T, Butler JRA, Williams LJ, Wanjura WJ, Hall A, Carter L, Carberry PS (2014) Taking complexity in food systems seriously: an interdisciplinary analysis. *World Development* **61**, 85–101. doi:10.1016/j.worlddev.2014.03.023

Fowler B, Dunn E (2014) *Evaluating Systems and Systemic Change for Inclusive Market Development: Literature Review and Synthesis*. Leveraging Economic Opportunities Report No. 3. USAID.

Gauri V, Woolcock M, Desai D (2013) Intersubjective meaning and collective action in developing societies: theory, evidence and policy implications. *Journal of Development Studies* **49**, 160–172. doi:10.1080/00220388.2012.700396

Glasbergen P (2007) Setting the scene: the partnership paradigm in the making. In *Partnerships, Governance and Sustainable Development: Reflections on Theory and Practice*. (Eds P Glasbergen, F Biermann, APJ Mol) pp. 1–28. Edward Elgar, Cheltenham.

Gómez MI, Barrett CB, Raney T, Pinstrup-Andersen P, Meerman J, Croppenstedt A, Carisma B, Thompson B (2013) Post-green revolution food systems and the triple burden of malnutrition. *Food Policy* **42**, 129–138. doi:10.1016/j.foodpol.2013.06.009

Gray B, Stites JP (2013) *Sustainability through Partnerships: Capitalizing on Collaboration.* Network for Business Sustainability, Pennsylvania State University.

Hall A (2011) *Putting Agricultural Research into Use: Lessons from Contested Visions of Innovation.* UNU-Merit Working Paper Series. UNU, Maastricht.

Hall A, Sulaiman VR, Clark N, Yogananda B (2003) From measuring impact to learning institutional lessons: an innovation systems perspective on improving the management of international agricultural research. *Agricultural Systems* **78**, 213–241. doi:10.1016/S0308-521X(03)00127-6

Hall A, Dijkman J, Sulaiman VR (2010) *Research into Use: Investigating the Relationship between Agricultural Research and Innovation.* UNU-MERIT Working Paper. UNU, Maastricht.

Hawkins R, Heemskerk W, Booth R, Daane J, Maatman A, Adekunle AA (2009) *Integrated Agricultural Research for Development (IAR4D).* Concept Paper for the Forum for Agricultural Research in Africa (FARA) Sub-Saharan Africa Challenge Programme (SSA CP). FARA, Accra, Ghana.

Hazlewood P (2015) *Global Multi-Stakeholder Partnerships: Scaling up Public-private Collective Impact for SDGs.* Background Paper No. 4. Independent Research Forum.

Horton D, Prain G, Thiele G (2009) *Perspectives on Partnership: A Literature Review.* Working Paper 2009–3. International Potato Center, Lima, Peru.

Hurlbert M, Gupta J (2015) The split ladder of participation: a diagnostic, strategic, and evaluation tool to assess when participation is necessary. *Environmental Science and Policy* **50**, 100–113. doi:10.1016/j.envsci.2015.01.011

Jerneck A, Olsson L, Ness B, Anderberg S, Baier M, Clark E, Hickler T, Hornborg A, Kronsell A, Lövbrand E, Persson J (2011) Structuring sustainability science. *Sustainability Science* **6**, 69–82. doi:10.1007/s11625-010-0117-x

Kates RW, Travis WR, Wilbanks TJ (2012) Transformational adaptation when incremental adaptations to climate change are insufficient. *Proceedings of the National Academy of Sciences of the United States of America* **109**(19), 7156–7161. doi:10.1073/pnas.1115521109

Klerkx L, Gildemacher P (2012) The role of innovation brokers in agricultural innovation systems. In *Agricultural Innovation Systems: An Investment Sourcebook.* pp. 221–230. World Bank, Washington DC.

Lahiri-Dutt K (2014) *Experiencing and Coping with Change: Women-headed Farming Households in the Eastern Gangetic Plains.* ACIAR Technical Report Series. ACIAR, Canberra.

Lansing JS (2003) Complex adaptive systems. *Annual Review of Anthropology* **32**(1), 183–204. doi:10.1146/annurev.anthro.32.061002.093440

Leach M (Ed) (2008) *Re-framing Resilience: A Symposium Report.* STEPS Working Paper 13. Social, Technological and Environmental Pathways to Sustainability Centre, Brighton, UK. <http://steps-centre.org/wp-content/uploads/Resilience.pdf>.

Leach M, Rockström J, Raskin P, Scoones I, Stirling AC, Smith A, Thompson J, Millstone E, Ely A, Arond E, Folke C, Olsson P (2012) Transforming innovation for sustainability. *Ecology and Society* **17**, 11. doi:10.5751/ES-04933-170211

Leitch A, Cundill G, Schultz L, Meek C (2015) Broaden participation. In *Principles for Building Resilience: Sustaining Ecosystem Services in Social-Ecological Systems.* (Eds R Biggs, M Schlüter, ML Schoon) pp. 201–225. Cambridge University Press, Cambridge.

Maru YT, Hall A, Banerjee O, Ison R, Butler JRA, Carberry P Appropriate mainstreaming of 'Theories of Change' approaches for development impacts. *Journal of Development in Practice.*

Millar J, Connell J (2010) Strategies for scaling out impacts from agricultural systems change: the case of forages and livestock production in Laos. *Agriculture and Human Values* **27**(2), 213–225. doi:10.1007/s10460-009-9194-9

Oteros-Rozas E, Martín-López B, Daw T, Bohensky E, Butler JRA, Hill R, Martin-Ortega J, Quinlan A, Ravera F, Ruiz-Mallén I, Thyresson M, Mistry J, Palomo I, Peterson G, Plieninger T, Waylen K, Beach D, Bohnet I, Hamann M, Hanspach J, Hubacek K, Lavorel S, Vilardy S (2015) Participatory scenario-planning in place-based social-ecological research: insights and experiences from 23 case studies. *Ecology and Society* **20**(4), 32. doi:10.5751/ES-07985-200432

Pahl-Wostl C (2009) A conceptual framework for analysing adaptive capacity and multi-level learning processes in resource governance regimes. *Global Environmental Change* **19**, 354–365. doi:10.1016/j.gloenvcha.2009.06.001

Parfitt T (2004) The ambiguity of participation: a qualified defence of participatory development. *Third World Quarterly* **25**, 537–555. doi:10.1080/0143659042000191429

Park SE, Marshall NA, Jakku E, Dowd AM, Howden SM, Mendham E, Fleming A (2012) Informing adaptation responses to climate change through theories of transformation. *Global Environmental Change* **22**, 115–126. doi:10.1016/j.gloenvcha.2011.10.003

Pelling M (2011) *Adaptation to Climate Change: From Resilience to Transformation*. Routledge, London.

Ramalingam B (2013) *Aid on the Edge of Chaos: Rethinking International Cooperation in a Complex World*. Oxford University Press, Melbourne.

Reed MS, Evely AC, Cundill G, Fazey I, Glass J, Laing A, Newig J, Parrish B, Prell C, Raymond C, Stringer LC (2010) What is social learning? *Ecology and Society* **15**, 4. http://www.ecologyandsociety.org/volXX/issYY/artZZ/

Reed MS, Podestá G, Fazey I, Geeson N, Hessel R, Hubacek K, Letson D, Nainggolan D, Prell C, Rickenbach MG, Ritsema C, Schwilch G, Stringer LC, Thomas AD (2013) Combining analytical frameworks to assess livelihood vulnerability to climate change and analyse adaptation options. *Ecological Economics* **94**, 66–77. doi:10.1016/j.ecolecon.2013.07.007

Rodima-Taylor D, Olwig MF, Chhetri N (2012) Adaptation as innovation, innovation as adaptation: an institutional approach to climate change. *Applied Geography (Sevenoaks, England)* **33**, 107–111. doi:10.1016/j.apgeog.2011.10.011

Roth CR, Grünbühel CM (2012) Developing multi-scale adaptation strategies: a case study for farming communities in Cambodia and Laos. *Asian Journal of Environment and Disaster Management* **4**(4) 425–446. doi:10.3850/S1793924012100055

Scoones I (2009) Livelihoods perspectives and rural development. *Journal of Peasant Studies* **36**(1), 171–196. doi:10.1080/03066150902820503

Skewes TD, Hunter C, Butler JRA, Lyne VD, Suadnya IW, Wise RM (2016) The Asset Drivers, Well-being Interaction Matrix (ADWIM): a participatory tool for estimating future impacts on ecosystem services and livelihoods. *Climate Risk Management* **12**, 69–82. doi:10.1016/j.crm.2015.08.001

Snowden M, Boone ME (2007) A leaders' framework for decision-making. *Harvard Business Review*, 1–9.

Stock P, Burton RJF (2011) Defining terms for integrated (multi-inter-trans-disciplinary) sustainability research. *Sustainability* **3**, 1090–1113. doi:10.3390/su3081090

Stone-Jovicich S, Butler JRA, McMillan L, Williams LJ, Roth C (2015) *Agricultural Research for Development in CSIRO: A Review of Principles and Practice for Impact*. CSIRO Agriculture, Brisbane.

Stringer LC, Dougill AJ, Fraser E, Hubacek K, Prell C, Reed MS (2006) Unpacking 'participation' in the adaptive management of social-ecological systems: a critical review. *Ecology and Society* **11**(2), 39. http://www.ecologyandsociety.org/vol11/iss2/art39/

Sugden F (2009) Neo-liberalism, markets and class structures on the Nepali lowlands: the political economy of agrarian change. *Geoforum* **40**, 634–644. doi:10.1016/j.geoforum.2009.03.010

Sugden F (2013) Pre-capitalist reproduction on the Nepal Tarai: semi-feudal agriculture in an era of globalisation. *Journal of Contemporary Asia* **1**, 27.

Sumberg J, Thompson J, Woodhouse P (2013) Why agronomy in the developing world has become contentious. *Agriculture and Human Values* **30**, 71–83. doi:10.1007/s10460-012-9376-8

UN (2014) *We Can End Poverty. Millennium Development Goals and Beyond 2015.* United Nations, New York. <http://www.un.org/millenniumgoals/poverty.shtml>.

UN (2015) *Open Working Group Proposal for Sustainable Development Goals.* United Nations, New York. <https://sustainabledevelopment.un.org/sdgsproposal>.

Vervoort JM, Thornton PK, Kristjanson P, Förch W, Ericksen PJ, Kok K, Ingram JSI, Herrero M, Palazzo A, Helfgott AES, Wilkinson A, Havlík P, Mason-D'Croz D, Jost J (2014) Challenges to scenario-guided adaptive action on food security under climate change. *Global Environmental Change* **28**, 383–394. doi:10.1016/j.gloenvcha.2014.03.001

Vogel I (2012) *Review of the Use of 'Theory of Change' in International Development.* UK Department for International Development, London.

Walker B, Gunderson L, Kinzig A, Folke C, Carpenter S, Schultz L (2006) A handful of heuristics and some propositions for understanding resilience in social-ecological systems. *Ecology and Society* **11**(1), 13. http://www.ecologyandsociety.org/vol11/iss1/art13/

Walker BH, Carpenter SR, Rockström J, Crépin A-S, Peterson GD (2012) Drivers, 'slow' variables, 'fast' variables, shocks, and resilience. *Ecology and Society* **17**, 30. doi:10.5751/ES-05063-170330

Williams LJ, Afroz S, Brown PR, Chialue L, Grünbühel CM, Jakimow T, Khan I, Minea M, Reddy V, Sacklokham S, Santoyo Rio E, Soeun M, Tallapragada C, Tom S, Roth CH (2015) Household types as a tool to understand adaptive capacity: case studies from Cambodia, Lao PDR, Bangladesh and India. *Climate and Development* online, 1–12. doi:10.1080/17565529.2015.1085362.

Wise RM, Butler JRA, Suadnya IW, Puspadi K, Suharto I, Skewes TD (2016) How climate compatible are livelihood adaptation strategies and development programs in rural Indonesia? *Climate Risk Management* **12**, 100–114. doi:10.1016/j.crm.2015.11.001

8

The co-construction of environmental (instream) flows and associated cultural ecosystem benefits

Rosalind H. Bark, Cathy J. Robinson, Sue E. Jackson and Karl W. Flessa

Management of water resources of major river basins, particularly of transboundary rivers (multistate or multination), is complex and has been the subject of much scholarship (Robinson et al. 2011; Bruns et al. 2005; Garrick 2015; Garrick et al. 2014; Bark et al. 2014). Issues surrounding institutional fragmentation, political contestation, scalar factors and challenges for collaboration are prominent in the literature on water governance and integrated water resource management. An emergent theme in the research is the importance of social, economic and especially cultural values associated with river systems and the development of frameworks that treat rivers and river management actions as part of a social, ecological and political system (Bark et al. 2016; Butler et al. 2013; Folke 2006; Jackson and Barber 2013; Lebel et al. 2013). Yet, in the growing number of studies concentrating on water governance, too little attention is given to the critical role of culture and cultural processes that are generated by interactions between humans and their environment and influenced by the dynamics of social experiences and interactions (Johnston et al. 2012).

People, society and water

Human cultures have developed around water, and their numerous religious systems and diverse poetic and musical forms revere rivers and celebrate symbols or rituals relating to water (Palmer 2015; Strang 1997). Social science has long been interested in interpreting the ways in which 'natural' land and waterscapes are socially constructed and invested with symbolic meaning. This work has been used to consider the social processes by which cultural landscape meanings are created and transmitted, to understand the profound differences in the ways that cultural groups define or construe the natural environment and the role these differences play in the cultural politics of natural resource management (Head 2000; Strang 1997; Jackson 2006). For example, at local levels in some countries, Indigenous cosmologies and environmental philosophies have endured colonial encounters and social science has shown the continuing impact of settler landscape ideologies on Indigenous rights and relationships with land/waterscapes (Jackson and Barber 2013; Robinson 2016). The multiple connections that social groups maintain with rivers represent particularly tight bonds between local social and environmental interactions because of water's essential role in sustaining human civilisations; indeed, in many cultures, the source of life itself is attributed to water (Klaver 2012).

Human–society relations with rivers are also evident and mediated through the formal and informal institutions established to make decisions about how water is managed, used and shared. There are many ways in which modes of resource governance are influenced by cultural processes, beyond the foundational role of religious belief and ontologies. Learned patterns of behaviour shape not only water access and use but also the local design of institutions, such as property relations, water sharing rules and payment for ecosystem services (PES). Rather than being an objective neutral process, economic valuation 'efforts to create the "right prices" for environmental assets entails political choices about which classes of people, in which geographic locations, will have access to natural resources and their benefits now and in the future' (McAfee and Shapiro 2010, p. 595; see also Robertson 2004). Formal and informal (laws, norms etc.) institutions that seek to internalise environmental impacts mediate human–water interactions and resource relations in a given context and locality and at particular moments in time. For decades, idealised design principles that ignore culture and cultural processes have been applied to water policy and management (Gupta et al. 2013; Margerum and Robinson 2015), affecting the extent to which they cater for the needs of local communities with diverse and rich connections to rivers and water as well as their own ways of valuing and governing water (Jackson and Altman 2009).

The ecosystem services framework, which is currently so popular, represents a promising option for tackling the challenge of bringing social and cultural processes to the fore in water management and governance (Jackson and Palmer 2012; Bark et al. 2015). The concept of ecosystem services describes the direct and indirect contributions of ecosystems to human well-being (MEA 2005). Cultural dimensions of ecosystems and the services they offer humans have become an important subset of this effort (Bark et al. 2015; Poe et al. 2014; Robinson et al. 2016; Stevens 2014). The cultural services offered by ecosystems have been defined to include non-material benefits, such as 'cultural diversity, spiritual and religious values, inspiration, aesthetic values, social relations, sense of place, cultural heritage values, recreation and ecotourism' (MEA 2005). Accordingly, policy agendas have attempted to acknowledge the cultural and social benefits that environments provide, and social scientists have considered a typology of cultural ecosystem benefits (Chan et al. 2012), how they can be mapped (Pert et al. 2015) and valued (Rolfe and Windle 2006; Zander et al. 2010), used to design and evaluate Indigenous PES agreements (Robinson et al. in review), operationalised into water management practice (Bark et al. 2014) and used to inform international standards for linking different types of knowledge to assess the state of the planet's biodiversity (Díaz et al. 2015).

A new goal emerging in the water planning literature is to expand understandings and applications of cultural ecosystem services, including attention to the implications that human–environment relationships are active, dynamic and often interdependent. Jackson and Palmer (2012) make a compelling argument for drawing on Indigenous practices and ethics of environmental care to provide a more holistic approach to PES. They argue that people *create* cultural ecosystem services through their care and action, that is, that nature is socialised and through the processes of socialising, people co-produce ecosystems services. This analysis, based on an Indigenous PES, questions the unsocialised nature supposed in the cascade where ecosystems and the ecosystem services that they provision supply human needs (Haines-Young and Potschin 2010). McAfee and Shapiro (2010) show how the ecosystem services paradigm separates nature and society (or culture) then reconnects these two categories of western thought. As a consequence, the conventional ecosystem services framework reductively constructs 'nature' as the sole provider of services so that it can be encompassed within 'economy'.

In this chapter, we argue that a fuller understanding of nature–society interrelationships and localised modes of resource governance is required to advance all major areas of integrated water resource management: from the assessment and amelioration of impacts of water infrastructure developments on local livelihoods and ways of life, to the design of allocation or water distribution/allocation mechanisms, regulations and incentives for aquatic conservation, public or citizen participation in catchment-scale governance institutions, and the effective restoration of degraded aquatic ecosystems. Specifically, we add to this innovative area of scholarship by tracing the expression of cultural ecosystem services over the lifecycle of a restoration event in the Colorado River Delta, the spring 2014 pulse flow. The pulse flow of 2014 was a negotiated restoration flow released down the usually dry riverbed south of the border between the US and Mexico: it was released in spring to recreate a snowmelt flood pulse of water timed to prepare (removing salts and stirring up) the riverbed and banks for spring germination.

The restoration flow was released in accordance with Minute 319 (Minute 319 2012)[1] to the 1944 Treaty governing water allocations between the US and Mexico. The significance of this restoration event is that it was the first international transboundary restoration event in the basin. The river had not reached the sea since 1998, which was a particularly wet year in the basin. The pulse flow was not an unplanned natural flow; it was the outcome of bilateral negotiation and it incorporated a monitoring protocol tasked to a binational team of biophysical scientists. The planned nature of the restoration flow made it an event suitable for study. Just as ecologists were interested in the ecological response to a water release, we were keen to observe and reflect on the social agency manifest in this pulse event and the human community responses to it. The pulse flow represents an interesting case study of the rich and understudied interrelationships between people and ecosystem services.

This chapter builds on previous work that tracked the ebb and flow of cultural ecosystem services as the environmental flow moved down the river and reached the sea (Bark *et al.* 2016) and on efforts to reconceptualise ecosystem service frameworks to recognise the interconnections between human–environment relationships and care (Díaz *et al.* 2015; Jackson and Palmer 2012; Robinson *et al.* 2016). The breakthrough comes from seeing nature as socially constructed and thus conceiving of cultural ecosystem services as generated or produced through human action, cognition and effect (Jackson and Palmer 2012). We apply this reconceptualisation to reveal the dynamics of human care and concern generated over the lifecycle of the pulse flow event. We find that revelation of nature–society interrelationships were (1) amplified over the course of the flow period, (2) that the event itself gave expression to a new vision or paradigm for negotiating and delivering the passage of amendments to treaty law between the US and Mexico that was reinvigorated when restoration water crossed international and state borders and pulsed past receiving communities, and (3) resulted in further post-event reflection on the social dimensions of environmental/instream flows and the future management of water resources in the basin to attain them.

A river revived

The Colorado River Basin is a transboundary river basin comprising four Upper Basin states (Wyoming, Utah, Colorado, New Mexico) and three Lower Basin states (Nevada, Arizona, California) in the US and two states (Baja California, Sonora) in Mexico. As the downstream

---

1   A Minute is the name given to an amendment to the 1944 Treaty (*1944 Treaty relating to the Utilization of Waters of the Colorado and Tijuana Rivers and of the Rio Grande*) that does not require full Congressional approval. As such, Minutes provide a more flexible and timely approach to modify Treaty arrangements.

nation, the delta ecosystem is entirely within Mexico. Water resources in the basin are highly regulated and allocation of water resources between the nations and the states is governed by a collection of treaties and rules, known as the 'law of the river'. A consequence of overallocation and the absence of water entitlements for the environment is that regular flows to the Gulf of California (which is in Mexico) ceased in 1960, as the upstream Lake Powell reservoir filled. Delta landscapes either dried out or were converted to irrigated agriculture. The last time that the Colorado River had reached the sea was in 1998, when upstream reservoirs were full – excess snowmelt was released to protected upstream dams and it had nowhere else to go but into the dry riverbed in Mexico and out to sea.

With a decade-long drought in the basin (2000 ongoing), 2014 was an inauspicious year for the first environmental flow to the delta. The drought has been dramatically portrayed with photographs of the white bathtub ring around Lake Mead, one of two main reservoirs in the basin, formed behind Hoover Dam. The low level of the reservoir, which has exposed the usually flooded sides of the lake, should not be alarming in itself, as a reservoir's function is to improve water supply reliability (Moy *et al.* 1986). This is imperative in a basin with interannual supply variability; however, the overallocation of the river's resources combined with little near-term prospect of filling the reservoir and growing demand, is a concern for water managers and communities reliant on it. A pulse flow, as per Minute 319, was set for spring sometime before 2017: the first environmental flow to cross the international boundary was released from Morelos Dam, a diversion dam that straddles the US–Mexico border, on 23 March 2014. During the next eight weeks (ceasing on 18 May 2014), 130 million m^3 (130 GL, 105 000 acre-feet) flowed downstream in a managed experiment, to water ecological restoration sites along the path of the once free-flowing Colorado River, to flush salts from the delta. Subsequent to the pulse flow, an additional 65 million m^3 will be delivered to river reaches and restoration sites up to the end of 2017 (Flessa *et al.* 2013). Groundwater levels at the restoration sites are sufficiently high that restored vegetation has a greater chance of survival. The goals of the pulse flow and base flows are to restore managed ecological sites, not the whole delta ecosystem, and there was no explicit goal for the pulse flow to reach the sea. Bilateral monitoring incorporates water quality, vegetation, wildlife and groundwater level measurements.

Using media reports for socio-cultural analysis

The pulse flow was the focus of local, national and international attention, celebration and commentary. We drew on media articles and blogs to uncover particular nature–society interrelationships associated with the Minute 319 pulse flow (Bark *et al.* 2016). Specifically, we collected media reports on the pulse flow between 30 December 2013 and 14 June 2014 using a daily Google News Alert involving the search terms 'pulse flow', 'Colorado River Delta', 'Colorado River' and 'Minute 319'. The coverage period included time before and after the pulse flow. We acknowledge that while in-person surveys provide immediate and intimate reflections, the need to apply a pragmatic cost-effective approach guided our choice to rely on media articles and blogs (McKellar *et al.* 2015). Analysis of 25 in-person interviews undertaken on 27 March 2014 – coincident with an official binational ceremony – can be found in Bark *et al.* (2016). For this chapter, the dataset consists of 153 English-language articles, opinion pieces and blogs.

To effectively manage the large number of individual articles and to address challenges about the rigour and credibility of social science research, we coded all the articles using content analysis software. Qualitative analysis tools like NVivo 10 or dedoose are often used to transform social data for analysis. Content analysis can reveal themes in data;

however, *ex ante* consideration on how and what to code is necessary. There are two main approaches to coding – grounded theory and frameworks. Grounded theory is a systematic approach to coding that is flexible to the data and themes that emerge (Strauss and Corbin 1990; Charmaz 2006). Using a framework is also a systematic approach to coding but it is inflexible and requires application of an appropriate framework.

For this research we used the framework approach, specifically adopting the cultural ecosystem benefits framework of Chan *et al.* (2012), which classifies cultural ecosystem benefits into 12 types: Activity, Aesthetic, Employment, Existence/bequest, Identity, Inspiration, Knowledge, Material, Option, Place/heritage, Social capital and cohesion, and Spiritual. We added a 13th coding option, Aspiration, as suggested by Bark *et al.* (2014). Each article was read and text was either left uncoded or coded for one or more cultural ecosystem benefit(s).

To further develop the explanatory power of the coding exercise, we also coded each article by its date: this date was then cross-referenced with the temporal phase of the pulse flow, that is Pre, Start, Peak, Connect(ion to the sea) and Post. In this way, each article was coded by a pulse flow phase and by the incidence of cultural ecosystem benefit(s) codes. Given the small number of codes for Material, we exclude it from this chapter.

The coded dataset enables us to test two hypotheses, H1 and H2, based on Johnston *et al.* (2012), where H_0 is the null hypothesis and H_1 is the alternative hypothesis. The outcome of hypothesis testing is that the evidence will either lead us to reject H_0 and accept H_1 because of sufficient evidence to support H_1, or not reject H_0 because of insufficient evidence to support H_1. The two hypotheses we test are shown below. Note that we are essentially testing whether the evidence supports H_1.

H_0 = cultural processes are not stimulated or generated by interactions between humans and their environment. That is, there is no sequencing of, or progression in, the presentation of cultural ecosystem benefits in the data/media reports. [H1]

H_1 = cultural processes are stimulated or generated by interactions between humans and their environment. That is, there is sequencing of, or progression in, the presentation of cultural ecosystem benefits in the data/media reports.

H_0 = cultural processes are not influenced by the social experiences and interactions triggered by the pulse flow. That is, the distribution of cultural ecosystem benefits is constant over the course of the flow. [H2]

H_1 = cultural processes are influenced by the social experiences and interactions triggered by the pulse flow. That is, cultural ecosystem benefits intensify during the pulse flow.

Once coded, the data allow us to examine two research questions. First, what evidence is there that cultural processes are stimulated or generated by interactions between humans and their environment; more particularly, will the presentation of cultural ecosystem benefits in the media reports be sequenced? And second, what evidence is there that cultural processes are influenced by the social experiences and interactions triggered by the pulse flow?

The ephemeral experience of the river returning

First we provide an example of each code type from the transcribed media reports. Activity: 'The pulse flow was made possible by a 2012 amendment to a 70-year old treaty between the US and Mexico'. Aesthetic: 'A once-lush area that has gone dry in recent decades with diversion of water upstream'. Aspiration: 'Now we are taking a major step to right the wrong that has been done to the Colorado River Delta'. Employment: 'This is a farm that's essentially in the riverbed, pumping groundwater. You see this type of farming on both sides of this stretch of the river'. Existence/bequest: 'For me, some of the most powerful experiences have come from visiting the same location twice – before the river got there, and then again after its arrival'. Identity: 'Happy to have the river back? ... Of course, amigo ... It's our name!' Inspiration: 'To be involved with something on this geographic magnitude is really quite remarkable'. Knowledge: 'Research teams from agencies, universities and environmental groups from both the US and Mexico will monitor the effects of this pulse, analysing the area before the flood, immediately following it and into the future'. Material: 'The river's freshwater estuary once nourished the sea for dozens of miles, creating an especially rich breeding zone'. Option: 'It's hard not to think of the lost potential of an ecosystem that once thrived'. Place/heritage: 'All this a long time ago was the river'. Social capital and cohesion: 'I think I underestimated the social impact ... This is aligning more than just river channels'. Spiritual: 'Just below the dam, at least, the river truly looked reborn', 'I try to keep myself open to the spiritual aspect of this work, instead of making it all about salinity and hydrology'.

The coding exercise provides count data for each cultural ecosystem benefit by pulse flow phase (see Table 8.1). It shows that the dominant cultural ecosystem benefits we coded for is Activity. Activity codes included negotiation, cooperation, celebration and implementing the pulse flow. The second most coded category was Aspiration, which coded language around the outcomes of the pulse flow, the meaning of the pulse flow and the opportunities to have one or many more pulse flow(s). The least coded category was Option, where we coded mention of charitable donations made by the general public to support ongoing restoration flows even when they are unlikely to ever visit the delta (i.e. option value). In terms of pulse flow phase, Connect, when the river and sea connected, had the most codes, followed by Peak, the peak water release of the pulse flow and Pre-pulse flow, the stories written before the pulse flow event. The initiation of the pulse flow release, Start, was the least coded time phase.

Table 8.1. Cultural ecosystem benefits by pulse flow phase, count data

	Pre	Start	Peak	Connect	Post	Total
Activity	78	54	73	74	61	**340**
Aesthetic	31	13	12	18	12	**86**
Aspiration	85	22	47	66	36	**256**
Employment	15	8	10	9	6	**48**
Existence/bequest	7	3	21	40	20	**91**
Identity	12	14	50	53	25	**154**
Inspiration	25	10	35	64	25	**159**
Knowledge	63	19	41	50	25	**198**
Option	12	2	6	6	7	**33**
Place/heritage	13	14	22	39	17	**105**
Social capital and cohesion	4	9	20	13	6	**52**
Spiritual	6	5	25	22	10	**68**
Total	**351**	**173**	**362**	**454**	**250**	

Note: Column total = total for each pulse flow phase. Row total = total for each cultural ecosystem category.

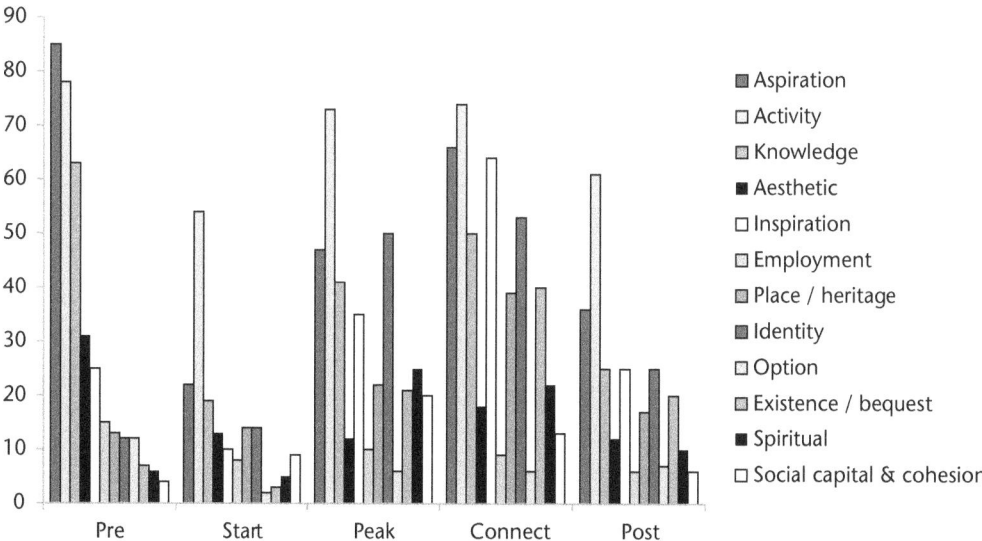

Fig. 8.1. Cultural ecosystem benefit types by pulse flow phase, ordered by Pre occurrence, count data.

Figures 8.1 and 8.2 display the information in Table 8.1 in two different ways that enable us to visually test our two hypotheses. Figure 8.1 shows the data with the cultural ecosystem benefits ordered not alphabetically, as in Table 8.1, but ordered by frequency in the Pre-pulse flow phase. The most coded cultural ecosystem benefit in this phase is Activity, and the least is Social capital and cohesion.

Figure 8.1 shows clearly that cultural processes are stimulated or generated by interactions between humans and their environment. That is, there is sequencing of, or progression in, the presentation of cultural ecosystem benefits in the data/media reports. In the Pre phase, a focus on typical outcomes to stimulate Action (i.e. negotiation and collabora-

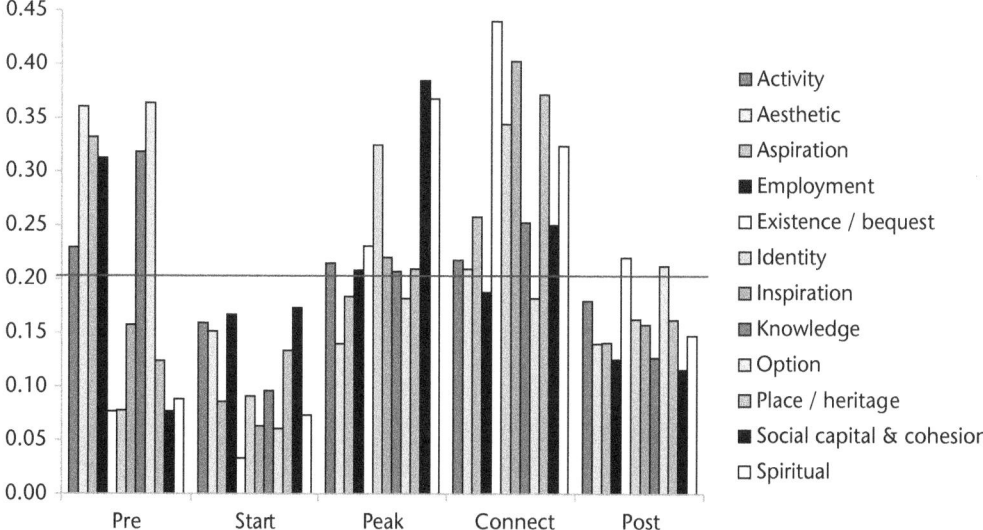

Fig. 8.2. Proportion of each cultural ecosystem benefit by pulse flow phase, ordered alphabetically. Note: any bar above/below 0.20 is over/underrepresented by pulse flow phase.

tion) and efforts of care highlight Knowledge that both underpinned the pulse flow and Knowledge that could be gained. Simultaneously, notions of Aspiration and Option value drove efforts by a dedicated group of conservation NGOs to envision a new paradigm in water sharing between the consumptive uses of water resources and setting aside water for nature. This was not a passive relationship but one of action, of envisioning, daring to change the status quo. In the Peak phase, Identity, and Social capital and cohesion were reflected. In the symbolic Connect phase, when the river joined the sea for one of the few times since the 1960s and the first time in 16 years, Existence/bequest and Inspiration were expressed, as were strong representations of Identity and Place/heritage. As the river returned to its full function as a river, people rejoiced, some identifying a Spiritual significance to this final moment in the sequence of events.

Figure 8.2 displays codes for each phase as a proportion of total codes for each cultural ecosystem benefit, for instance, coding for Activity in the Peak phase represents 21% or 0.21 of the total codes for Activity, i.e. $73/340 = 0.21$.

Figure 8.2 partially shows that cultural processes are influenced by the social experiences and interactions triggered by the pulse flow. That is, cultural ecosystem benefits intensify during the pulse flow. We say 'partially' as Fig. 8.2 does not show a build-up in all cultural ecosystem benefits with flow. Rather, it shows a build-up of place-based, social and inspiration cultural ecosystem benefits that may be related to the excitement generated by the event – a demonstration of the immediate effect of the restoration flow, i.e. interaction with the pulse flow and with others experiencing the pulse flow.

Ecosystem services embedded in sociocultural systems

Our goal in this research was to show the interwoven, socio-cultural basis of ecosystem service construction that is conventionally neglected by the cultural ecosystem services approach. Humans are not just encountering nature but radically altering environments, both consciously and inadvertently, and in doing so are creating new value from the social occasion of planning, witnessing and interpreting a flow event. Prior to the pulse flow the Colorado River became the focus of human aspiration and debate – as well as fear that it was poorly timed during a drought. The data show that when it actually flowed people rejoiced, remembered, celebrated, bathed, affirmed their connections, and dreamed. Social and ecological relationships intersected and interacted to prepare and respond to the river being given the water to again flow into the sea, building on the excitement of the moment. In so doing, people *and* the river generated more intense and diverse values. The aspiration to revitalise a river was grounded in ideals about ecological conservation and fairness; perhaps surprisingly, only when water flowed was it appreciated that another dimension was evident, namely the human dimension. The return of the river's flow enhanced interaction with the river and with each other as those involved in the negotiation of the pulse flow – commentators, journalists and locals – were struck with awe, jubilation and hope. The dataset generated from media articles and blogs provided a means to view this interaction and co-production of value; we see that coding for cultural ecosystem benefits works and we see how cultural ecosystem benefits change in nature/type and intensity. The coding by time or pulse flow phase facilitated the testing of research questions about the dynamic and socially constructed nature of cultural ecosystem benefits and of broader themes such as interdependence, sequencing, engagement and reflection.

Restoration as an event is a good case study to test concepts of generation and dynamics in cultural ecosystem services (Johnston *et al.* 2012). We did this by focusing on two research questions. The data suggest that cultural process are generated by interactions between humans and their environment, that there is sequencing of, or progression in, the presentation of types of cultural ecosystem benefits, and that cultural processes are influenced by the dynamics of social experiences and interactions triggered by a flow event. Akin to Urquhart and Acott's (2014, p. 15) contention that fishing practice in Cornwall puts in motion events that manifest sense of place and identity on land, the pulse flow restoration event precipitated changes in the riverscape that in turn co-construct identities and cultural ecosystem benefits (or values) such as place attachment. The data also support Jackson and Palmer's (2012) reasoning about the interconnections between human–environment relationships and care. There is clear evidence of the effect of a reborn river on the co-produced cultural ecosystem benefits expressed in media reports. The Colorado River is a socialised river – nations, locals and commentators around the world engaged with different phases of the flow. This engagement peaked when the flow peaked, then again when it connected to the sea. The results suggest that human aspiration and action were determinative for the delta environment, and that these actions and aspirations in turn prompted a suite of other human–environment relationships and interpretations of the event. Aspirations for future management of this river were embedded in narratives revealed through the media analysis.

This final reflection phase returns to the role of human care and action needed for the next pulse flow. The (first) transboundary flow in 2014 was shepherded through bilateral negotiations by a determined group of NGOs whose aspirational goals were to renew a river. They were supported by scientists and scientific information that showed that the delta ecosystem was not dead, but dormant, and that it had responded in the past to previous floods or excess flows (Glenn et al. 2007, 2013). The connection of river and sea, though not an ecological goal of the pulse flow, was not only symbolic of the river reaching its final destination but was reflected in media coverage as symbolic of a new paradigm in people's relationships with the river that would necessarily release a community consensus and perhaps the political will to change future management of the river along its entire journey.

The method used, content analysis, is reproducible and rigorous. Our sampling of media articles was transparent; however, the approach relied on the pulse flow being a big media event. The short timeframe of the pulse flow might also have contributed to ongoing media interest. Thus the replicability of this method relies on other restoration case studies being newsworthy and with set start and end dates. Furthermore, our analysis would be enriched with inclusion of Spanish-language media, i.e. Mexican. We excluded Material cultural ecosystem benefits from our analysis (e.g. mention of irrigation output or commercial fishing) as it was rarely mentioned in our dataset. This might be because the volume of water in the environmental flow was deemed so small that its effects would be negligible on irrigation output, or because the trade-off with irrigation output was implicit and therefore not mentioned, or the water delivered was viewed as having a zero opportunity cost (Bark et al. 2014). Whatever the explanation, the low profile of what can be a contentious trade-off may have helped in the illumination of what we contend is a socially constructed riverscape and delta.

Conclusion: multifaceted ecosystem services

The pulse flow restored and strengthened awareness of the cultural aspects of ecosystem services. The research highlights the interactive and dynamic nature of cultural ecosystem benefits as communities, NGOs and scientists built knowledge, negotiated the pulse flow release, engaged in political coalitions and created the consensus to realise this huge restoration event, revealing an active relationship with nature and a suite of subsequent cultural ecosystem benefits. This analysis points to the limitations of considering nature as a 'service provider' and of ignoring the diverse and important ways in which humans contribute to socio-ecological processes and functions. We note the critical contribution of social scientists to the thinking around the ecosystem service framework that is so influential in global environmental management (Dempsey and Roberston 2012; Kosoy and Corbera 2010). This includes the growing concern that the ecosystem service framework considers humans as detached and insulated from ecosystem processes, which, as Armsworth et al. (2007, p. 1384) note, is both outdated and dangerous.

Social scientists have offered pragmatic approaches to improve the thinking and application surrounding ecosystem services that pay appropriate attention to the active, reflexive human–environment relationships that 'service' each other. The example of the pulse flow in the Colorado River highlights the nuances involved in this interactive and reflexive relationship. By reviving, returning, reformulating and re-enchanting these services, the restoration event highlights the immediacy of this interconnected relationship, one that was dormant and awaiting restoration. It acts as a reminder or challenge for scientists, researchers and policy-makers to look for and value these kinds of relationships in other contexts as they explore ecosystem services.

Acknowledgments

We thank the Distinguished Visiting Scientist program of CSIRO, the National Environmental Science Programme, and the CSIRO Payne-Scott Award. This project received funding from the European Union's Horizon 2020 research and innovation program under the Marie Skłodowska-Curie grant agreement No. 659449.

References

Armsworth PR, Chan KMA, Daily GC, Ehrlich PR, Kremen C, Ricketts TH et al. (2007) Ecosystem-service science and the way forward for conservation. *Conservation Biology* **21**, 1383–1384. doi:10.1111/j.1523-1739.2007.00821.x

Bark RH, Frisvold GB, Flessa KW (2014) The role of economics in transboundary restoration water management in the Colorado River Delta. *Water Resources and Economics* **8**, 43–56. doi:10.1016/j.wre.2014.10.006

Bark RH, Barber M, Jackson S, Maclean K, Pollino C, Moggridge B (2015) Operationalising the ecosystem services approach in water planning: a case study of indigenous cultural values from the Murray-Darling Basin, Australia. *International Journal of Biodiversity Science, Ecosystem Services and Management* **11**(3), 239–249.

Bark RH, Robinson CR, Flessa K (2016) Tracking cultural ecosystem services: water chasing the Colorado River restoration pulse flow. *Ecological Economics* **127**, 165–172. doi: 10.1016/j.ecolecon.2016.03.009

Bruns BR, Ringler C, Meinzen-Dick RS (Eds) (2005) *Water Rights Reform: Lessons for Institutional Design*. International Food Policy Research Institute, Washington DC. <http://www.ifpri.org/pubs/books/oc49/oc49.pdf>.

Butler JRA, Wong GY, Metcalfe DJ, Honzak M, Pert PL, Rao N, van Grieken ME, Lawson T, Bruce C, Kroon FJ, Brodie JE (2013) An analysis of trade-offs between multiple ecosystem services and stakeholders linked to land use and water quality management in the Great

Barrier Reef, Australia. *Agriculture, Ecosystems and Environment* **180**, 176–191. doi:10.1016/j.agee.2011.08.017

Chan KMA, Satterfield T, Goldstein J (2012) Rethinking ecosystem services to better address and navigate cultural values. *Ecological Economics* **74**, 8–18. doi:10.1016/j.ecolecon.2011.11.011

Charmaz K (2006) *Constructing Grounded Theory: A Practical Guide through Qualitative Analysis*. Sage, London.

Dempsey J, Robertson MM (2012) Ecosystem services: tensions, impurities, and points of engagement within neoliberalism. *Progress in Human Geography* **36**(6), 758–779. doi:10.1177/0309132512437076

Díaz S, Demissew S, Carabias J, Joly C, Lonsdale M, Ash N, Larigauderie A, Adhikari JR, Arico S, Báldi A, Bartuska A, Baste IA, Bilgin A, Brondizio E, Chan KMA, Figueroa VE, Duraiappah A, Fischer M, Hill R, Koetz T, Leadley P, Lyver P, Mace GM, Martin-Lopez B, Okumura M, Pacheco D, Pascual U, Pérez EW, Reyers B, Roth E, Saito O, Scholes RJ, Sharma N, Tallis H, Thaman R, Watson R, Yahara T, Hamid ZA, Akosim C, Al-Hafedh Y, Allahverdiyev R, Amankwah E, Asah ST, Asfaw Z, Bartus G, Brooks LA, Caillaux J, Dalle G, Darnaedi D, Driver A, Erpul G, Escobar-Eyzaguirre P, Failler P, Fouda AMM, Fu B, Gundimeda H, Hashimoto S, Homer F, Lavorel S, Lichtenstein G, Mala WA, Mandivenyi W, Matczak P, Mbizvo C, Mehrdadi M, Metzger JP, Mikissa JB, Moller H, Mooney HA, Mumby P, Nagendra H, Nesshover C, Oteng-Yeboah AA, Pataki G, Roué M, Rubis J, Schultz M, Smith P, Sumaila R, Takeuchi K, Thomas S, Verma M, Yeo-Chang Y, Zlatanova D (2015) The IPBES Conceptual Framework: connecting nature and people. *Current Opinion in Environmental Sustainability* **14**, 1–16. doi:10.1016/j.cosust.2014.11.002

Flessa KW, Glenn EP, Hinojosa-Huerta O, de la Parra-Renteria CA, Ramirez-Hernandez J, Schmidt JC, Zamora-Arroyo F (2013) Flooding the Colorado River delta: a landscape-scale experiment. *EOS* **94**, 485–486. doi:10.1002/2013EO500001

Folke C (2006) Resilience: the emergence of a perspective for social-ecological systems analyses. *Global Environmental Change* **16**, 253–267. doi:10.1016/j.gloenvcha.2006.04.002

Garrick D (2015) *Water Allocation in Rivers under Pressure: Water Trading, Transaction Costs and Transboundary Governance in the Western US and Australia*. Edward Elgar, Cheltenham.

Garrick DG, Anderson G, Connell D, Pittock J (Eds) (2014) *Federal Rivers: Managing Water in Multi-layered Political Systems*. Edward Elgar, Cheltenham.

Glenn EP, Flessa KW, Cohen MJ, Nagler PL (2007) Just add water and the Colorado River still reaches the sea. *Environmental Management* **40**, 1–6. doi:10.1007/s00267-006-0070-8

Glenn EP, Flessa KW, Pitt J (2013) Restoration potential of the aquatic ecosystems of the Colorado River delta, Mexico: introduction to special issue on 'Wetlands of the Colorado River Delta'. *Ecological Engineering* **59**, 1–6. doi:10.1016/j.ecoleng.2013.04.057

Gupta J, Pahl-Wostl C, Zondervan R (2013) 'Global' water governance: a multi-level challenge in the Anthropocene. *Current Opinion in Environmental Sustainability* **5**, 573–580. doi:10.1016/j.cosust.2013.09.003

Haines-Young R, Potschin M (2010) The links between biodiversity, ecosystem services and human well-being. In *Ecosystem Ecology: A New Synthesis*. (Eds D Raffaelli, C Frid) pp. 110–139. Cambridge University Press, Cambridge.

Head L (2000) *Cultural Landscapes and Environmental Change*. Oxford University Press, London.

Jackson S (2006) Compartmentalising culture: the articulation and consideration of Indigenous values in water resource management. *Australian Geographer* **37**(1), 19–31. doi:10.1080/00049180500511947

Jackson S, Altman J (2009) Indigenous rights and water policy: perspectives from tropical northern Australia. *Australian Indigenous Law Review* **13**(1), 27–48.

Jackson S, Barber M (2013) Recognition of Indigenous water values in Australia's Northern Territory: current progress and ongoing challenges for social justice in water planning. *Planning Theory and Practice* **14**(4), 435–454. doi:10.1080/14649357.2013.845684

Jackson S, Palmer L (2012) Modernising water: articulating custom in water governance in Australia and Timor-Leste. *International Indigenous Policy Journal* **3**(3). doi:10.18584/iipj.2012.3.3.7.

Johnston BR, Hiwasaki L, Klaver IJ, Ramos Castillo A, Strang V (Eds) (2012) *Water, Cultural Diversity and Global Environmental Change: Emerging Trends, Sustainable Futures?* UNESCO/Springer SBM, Paris.

Klaver C (2012) Culture, capital and representation. *Economic History Review* **65**(3), 1207–1209. doi:10.1111/j.1468-0289.2011.00644_29.x

Kosoy N, Corbera E (2010) Payments for ecosystem services as commodity fetishism. *Ecological Economics* **69**, 1228–1236. doi:10.1016/j.ecolecon.2009.11.002

Lebel L, Nikitina E, Pahl-Wostl C, Knieper C (2013) Institutional fit and river basin governance: a new approach using multiple composite measures. *Ecology and Society* **18**(1), 1. doi:10.5751/ES-05097-180101

Margerum R, Robinson CJ (2015) Limitations of collaborative partnerships for sustainable water management. *Current Opinion in Environmental Sustainability* **12**, 53–58. doi:10.1016/j.cosust.2014.09.003

McAfee K, Shapiro EN (2010) Payments for ecosystem services in Mexico: nature, neoliberalism, social movements, and the state. *Annals of the Association of American Geographers* **100**(3), 579–599. doi:10.1080/00045601003794833

McKellar L, Bark RH, Watson I (2015) Agricultural transition and land use change: considerations in development of irrigated enterprises in rangelands of northern Australia. *Rangeland Journal* **37**(5), 445–457. doi:10.1071/RJ14129

MEA (Millennium Ecosystem Assessment) (2005) *Ecosystems and Human Well-being: Synthesis*. Island Press, Washington DC.

Minute 319 (2012) *Amendment to 'Treaty for the Utilization of Waters of the Colorado and Tijuana Rivers and of the Rio Grande'*. Interim international cooperative measures in the Colorado River Basin through 2017 and extension of Minute 318 cooperative measures to address the continued effects of the April 2010 earthquake in the Mexicali Valley, Baja California, signed in 1944, Minute 319 signed in November, 2012. <http://www.ibwc.gov/Files/Minutes/Minute_319.pdf>.

Moy W-S, Cohon JL, ReVelle CS (1986) A programming model for analysis of the reliability, resilience, and vulnerability of a water supply reservoir. *Water Resources Research* **22**(4), 489–498. doi:10.1029/WR022i004p00489

Palmer L (2015) *Water Politics and Spiritual Ecology: Custom, Environmental Governance and Development*. Routledge, London.

Pert PL, Hill R, Maclean K, Dale A, Rist P, Talbot LD, Tawake L, Schmider J (2015) Mapping cultural ecosystem services with rainforest aboriginal peoples: integrating biocultural diversity, governance and social variation. *Ecosystem Services* **13**, 41–56. doi:10.1016/j.ecoser.2014.10.012

Poe MR, Norman KC, Levin PS (2014) Cultural dimensions of socioecological systems: key connections and guiding principles for conservation in coastal environments. *Conservation Letters* **7**, 166–175.

Robertson M (2004) The neoliberalization of ecosystem services: wetland mitigation banking and problems in environmental governance. *Geoforum* **35**, 361–373. doi:10.1016/j.geoforum.2003.06.002

Robinson CJ (2016) Hunting for country and culture: the challenges surrounding Indigenous collaborative partnerships on the coast of northern Australia. In *The Challenges of Collaboration in Environmental Governance: Barriers and Responses*. (Eds R Margerum, CJ Robinson) pp. 355–368. Edward Elgar, Cheltenham. doi:10.4337/9781785360411.0002

Robinson CJ, Margerum RD, Koontz TM, Mosely C, Lurie S (2011) Policy-level collaboratives for environmental management at the regional scale: lessons and challenges from Australia and the United States. *Society and Natural Resources* **24**, 849–859. doi:10.1080/08941920.2010.487848

Robinson CJ, James G, Whitehead PJ (2016) Negotiating Indigenous benefits from payment from ecosystem (PES) schemes. *Global Environmental Change* **28**, 21–29.

Rolfe J, Windle J (2006) Valuing Aboriginal cultural heritage across different population groups. In *Choice Modelling and the Transfer of Environmental Values*. (Eds J Rolfe, J Bennett) pp. 216–244. Edward Elgar, Cheltenham.

Strang V (1997) *Uncommon Ground: Cultural Landscapes and Environmental Values*. Berg, Oxford.

Stevens S (Ed) (2014) *Indigenous Peoples, National Parks and Protected Areas*. University of Arizona Press, Tucson.

Strauss A, Corbin J (1990) *Basics of Qualitative Research: Grounded Theory Procedures and Techniques*. Sage, London.

Urquhart J, Acott T (2014) A sense of place in cultural ecosystem services: the case of Cornish fishing communities. *Society and Natural Resources* **27**(1), 3–19. doi:10.1080/08941920.2013.820811

Zander KK, Garnett S, Straton A (2010) Trade-offs between development, culture and conservation: willingness to pay for tropical river management among urban Australians. *Journal of Environmental Management* **91**, 2519–2528. doi:10.1016/j.jenvman.2010.07.012

9

Dipping in the well: how behaviours and attitudes influence urban water security

Anneliese Spinks, Kelly Fielding, Aditi Mankad, Rosemary Leonard, Zoe Leviston and John Gardner

Water security poses one of the greatest challenges to urban life as water scarcity not only threatens the desires and aspirations of society, but also its survival. The Australian population is relatively small compared to that of other land masses across the globe, but the particular combination of low and variable rainfall combined with rapidly expanding urban areas means that Australian society frequently faces the problem of insufficient water supply. In particular, a prolonged and widespread drought affecting the southern and eastern states during the first decade of the 21st century (now known as the 'Millennium drought') had a profound impact upon the way Australian people perceive the role of water in their cities and the wider environment.

During the Millennium drought, the threat of depleted water supplies hovered menacingly, and a new era of water planning and efficiency began. This in turn drove a comprehensive research agenda exploring how best to ensure sufficient water availability for the future. Understanding how people use, perceive and relate to water became recognised as an essential part of this agenda, particularly given that individual habits, knowledge and core values underlie the ways in which people interact with technology and their environment.

Technological solutions for urban centres have been essential to ensuring water security and often involve large-scale infrastructure projects such as the construction of dams, pipelines, desalination plants and water treatment facilities. Newer technologies involving the treatment and reuse of stormwater and wastewater have also been proffered as part of the technological mix, while smaller-scale infrastructure solutions include the installation of rainwater tanks and greywater reuse schemes at household and community levels. Within homes and businesses, water-efficient devices such as low-flow shower heads, water-efficient dishwashers and washing machines, and dual-flush toilet systems have been widely encouraged in a bid to increase the efficiency of residential water use.

The success of these technological solutions does not rest entirely upon the reliability of the engineering component, but also upon the way in which the technology is perceived, embraced and used by individuals. Indeed, the public reaction to innovation in water infrastructure can influence outcomes in ways not anticipated by decision-makers. This holds true across the spectrum of water supply and demand. For example, governments have been poised to allow treated wastewater to flow through pipelines into households for drinking and bathing, yet have been stymied by vocal opposition and condemnation (Dolnicar and Hurlimann 2009; Price *et al.* 2012); innovative water engineers have been baffled that the performance of water-saving appliances installed throughout homes and businesses has failed to match modelled and projected savings (Inman and Jeffrey 2006); and household rainwater tanks have the potential to transform into fetid breeding grounds for mosquitoes while home-owners remain unaware, through lack of interest or knowledge.

Societal reactions to water-related issues and technology are driven by community attitudes and beliefs. Our interaction with water is a vital and fundamental part of the human existence, and is often driven by deep-seated emotional reactions that may, at the extreme, override practical and rational considerations. These deeply embedded responses are compounded by the emergence of modern societal desires for cleanliness and purity, as manifested in sparkling homes and lush green gardens (Shove 2003). Hence, an understanding of the way in which individuals think about and use water is essential for the success of efforts to ensure urban water security.

In this chapter, we will describe a comprehensive body of social research undertaken by CSIRO to investigate people's attitudes to, acceptance of and behaviour around urban water supply and demand. We will illustrate the role of people's attitudes and beliefs in shaping urban water security through a series of case studies designed to highlight the major findings and recommendations of our work across different water augmentation sources and at different scales of the water supply system.

The urban water supply

Water can be supplied for residential and commercial purposes within Australian towns and cities in many ways. Sources of water include naturally occurring rivers and lakes, constructed reservoirs, underground aquifers and rainwater harvested from roofs and stored in tanks. In addition, there are water reuse options that draw on stormwater and wastewater, and salt can be removed from seawater to make it potable. Water from different sources requires different types and intensity of treatment to ensure it is suitable for its intended use. Supply networks can also be configured so that they provide water that is fit-for-purpose. For example, dual pipe systems provide water intended for potable (drinking and cooking) applications separately from water intended for non-potable uses such as outdoor irrigation and clothes washing. Wastewater systems can also be constructed so that they separate blackwater from greywater, making subsequent water reuse treatment a less intensive process.

Securing supply: an Adelaide case study

Community preferences for the way in which water is supplied are shaped by the knowledge, attitudes and expectations of individuals and communities in relation to the associated environmental, social and economic costs. This is illustrated in a case study exploring the optimal mix of water sources to secure the Adelaide metropolitan water supply. Adelaide has endured a long history of water scarcity, and its traditional catchments within the Mount Lofty Ranges and the Murray–Darling Basin have suffered marked degradation over decades. Hence, water authorities needed to consider a range of water supply options to meet the simultaneous challenges of a growing population and industry demands.

Proposed options for securing the water supply included increasing the amount of water drawn from traditional sources (the River Murray and Mount Lofty Ranges), expanding the operations of the recently constructed desalination plant, reusing stormwater and wastewater, and increasing reliance on water sourced from rainwater tanks and groundwater systems. The state government considered that the best scenario would involve a combination of these water sources, and a decision needed to be made on the 'optimal mix'. Authorities were unsure, however, how the community would respond to decisions made to secure the water supply, and wanted to establish how public sentiment would accord with best engineering, financial and environmental practice.

In 2013, we engaged with community members in a series of focus groups during which photographic cards depicting the different options were employed to stimulate discussion and elicit reactions to the various water supply scenarios described above (Cuddy et al. 2014). The discussion included the potential advantages and disadvantages of each source (including realistic financial and environmental costs). Following the discussion, participants were asked to vote for their preferred 'mix' of water sources to supply Adelaide with adequate water to meet projected demands to 2050.

Key findings

The focus groups revealed a high degree of community support for supplementing the water supply with treated stormwater, wastewater and captured rainwater, for both potable and non-potable uses. These water sources were generally favoured over the options of drawing more water from the River Murray or expanding desalination operations, and were perceived to offer environmental benefits associated with protecting natural ecosystems (e.g. the Murray–Darling Basin and the coastal water systems). There was, however, a strong need for reassurance that authorities would manage the systems adequately, with some degree of mistrust expressed about existing water management arrangements (Cuddy et al. 2014). Although water desalination was not particularly favoured for providing water, community members conveyed the sense that, because a desalination plant had already been built, it would be a useful means for extra water, especially in times of water scarcity. Participants expressed concerns regarding the financial and energy-related costs of developing and maintaining water-related technology, and the lack of adequate community engagement. Thus, it appears more likely that a positive community response will prevail if authorities are able to demonstrate efficient usage of existing infrastructure, water, money and energy (Cuddy et al. 2014).

Community responses to water reuse

Factors related to either the acceptance of, or objection to, the reuse of stormwater and wastewater have been explored within several Australian communities. Stormwater is the surface runoff from urban areas, collected from a network of drains and diverted to waterways which eventually flow into the ocean. Wastewater is the sewage output discarded from homes and other private and commercial premises. While stormwater is normally untreated or minimally treated, wastewater undergoes a series of physical, chemical and biological cleansing processes before being discharged into the environment. Both stormwater and wastewater have the potential to be recycled for potable and non-potable applications. For this to occur, additional treatment processes must be conducted to remove potential contaminants which, depending on the source, can include various combinations of human and animal waste, soil particles, heavy metals, organic compounds, pharmaceuticals, oils and grease.

Recycling wastewater in south-east Queensland

In 2008, nine focus groups were conducted with 99 Queensland community members to ask their views regarding the potential addition of treated wastewater to Wivenhoe Dam, the main reservoir for the region's supply of drinking water (Alexander *et al.* 2008). The terminology adopted by the Queensland government to refer to the treated wastewater was 'purified recycled water' (PRW). The community consultation occurred at a time when dam levels were at historic lows due to the prolonged drought, and the community had faced unprecedented levels of water restrictions. Dusty cars and wilted shrubbery had become the aesthetic norm in a region more accustomed to tropical greenery. Thus, a sense of inevitability surrounded the impending decision to add PRW to the existing water supply, with most community members seemingly resigned to, if not accepting, the proposed course of action.

Following the focus groups, a series of telephone-based surveys was implemented to assess changes in people's attitudes towards the proposed PRW scheme over time (Price *et al.* 2010). In total, 1618 individual surveys were conducted in four waves between November 2007 and December 2009, with a significant proportion of individuals participating in more than one survey (approximately one-third of the total number of surveys were conducted with an individual who had completed one or more previous surveys). The final wave of surveys (conducted in December 2009) occurred after a period of significant rainfall had boosted dam levels and prompted the Queensland government to announce a decision that PRW would be introduced to the drinking water supply only if the dam levels fell below 40% again.

Key findings

Individuals participating in the focus groups generally conveyed a need for caution and the sense that support for the decision to introduce PRW to the drinking water supply was highly contingent upon the deepening supply crisis. Although a high level of trust in science and technology was expressed, there was still concern surrounding potential system risk and the likelihood of significant health impacts resulting from mismanagement of the system (i.e. the prospect of inadequately treated water infiltrating and contaminating the main water supply.) The perception of risk was connected to issues of trust, fairness and emotion, with emotion – the so-called yuck factor – being a pervasive influence that was not necessarily overridden by the availability of information and scientific reassurance. Hence, despite acknowledging the presence of contaminants in natural environments, some residents maintained that drinking treated wastewater was 'a violation of the natural order' and that no degree of treatment could ever render the water 'clean enough'. For some people, emotional responses of anxiety, fear, disgust and resentment were not assuaged by scientific flowcharts and assurances of technological safeguards.

Across the four telephone surveys, support for the PRW scheme was relatively high, with at least 70% of participants reporting that they were willing to drink the recycled water. However, most people indicated they would prefer not to drink the PRW if a viable alternative were available. Clear differences between supporters, opponents and unsure responders of the PRW scheme persisted across the four survey waves. Supporters generally trusted that the government would use the best science to underpin the scheme and would ultimately provide a safe and healthy water supply. They also considered the decision-making processes to be fair, while opponents and those unsure about the scheme did not think these processes had been fairly undertaken. Support for the scheme was substantially threatened by the potential for negative environmental impacts, with only 26% of respondents reporting that they would be supportive if the environment would be harmed.

In summary, the results suggest that it is possible to achieve acceptance among the majority of community members for potable water reuse schemes; however, there may be some challenges involved in maintaining that support. Trust in government to deliver a safe and healthy water supply needs to be developed, as do efforts to ensure that the scheme is perceived to be fair. It must also be acknowledged that a minority of community members will not accept the scheme under any circumstances, and that pockets of resistance will persist regardless of efforts at engagement and information provision.

Before treated wastewater could be added to the existing water supply, rain started to fall. Wivenhoe Dam subsequently overfilled, causing widespread flooding, and the catchment dams have remained at near capacity since. Not surprisingly, talk of supplementing the water supply with PRW has, for now, disappeared from the community dialogue.

Wastewater in Perth

Perth, the capital of Western Australia, has had considerable water supply challenges over the years, exacerbated by a drying climate and surges in population during mining upturns. In accounting for projected reduced rainfall coupled with a growing population, the Water Corporation of Western Australia estimated an additional 120 GL of water per annum would be required by 2030 (an additional 40% compared to that used in 2009). Managed aquifer recharge (MAR; also known as groundwater replenishment) has been approved by the Western Australian government as a means for securing Perth drinking water supplies. MAR also helps maintain the environmental sustainability of the aquifer Perth relies on for ~60% of its drinking water supply. Prior to the approval, a series of trials involving MAR for both non-potable and potable uses was carried out in urban Perth and in nearby regional areas as part of a major multidisciplinary study funded by the Western Australian Premier's Water Foundation. Part of the study involved an analysis of community attitudes in Perth to using treated MAR wastewater for potable purposes (Leviston *et al.* 2009). A further component of the study investigated risk perceptions associated with a proposed managed aquifer recharge scheme using QMethodology to compare decision-making frameworks of lay community with 'technical expert' assumptions about the lay community (Leviston *et al.* 2013).

Key findings

Findings from a behavioural modelling survey of 500 Perth householders suggested that, while over half the respondents intended to support an MAR scheme in Perth, a significant proportion (one-fifth) did not have strong convictions about the scheme, expressing moderate responses in relation to intended behaviour towards the scheme. Roughly one-quarter of respondents indicated refusal to support the scheme. While there were no statistically significant differences based on education levels, income levels, family unit and age, males' behavioural intentions towards the scheme were significantly more positive than were females'. Emotion and subjective norms (the extent to which participants' perceived important others supported the scheme) were found to have the strongest direct influence on intended behaviour in relation to an indirect potable MAR scheme. Fairness, trust, and perceived health and system risks also had significant influences. Importantly, self-reported knowledge failed to contribute significantly to the prediction of intended behaviour.

In the QMethodology decision-making study, despite a general consensus in the literature about the salience of emotion-based assessments for community in making decisions about risk, particularly where wastewater is concerned, emotion failed to play a significant role in the communities' decision-making frameworks. This suggests that the link between emotion and risk is complex, and that emotion-based assessments go beyond the 'yuck factor' to include other more contextual considerations and assessments, including governance accountability and regulatory concerns, holistic water supply management, and a desire for participatory decision-making – all of which featured prominently in decision-making frameworks. These findings were in sharp contrast to experts' assumptions about what the community considered important. Technical experts' dominant depiction of community members is inaccurately caricatured as relatively non-accepting, largely emotion-focused and driven, focused on health concerns, and reflecting a lack of trust and confidence in scientific, policy and management processes. This research highlights that there can be a mismatch between how communities actually respond to water reuse compared to how experts think they are responding.

Stormwater recycling in Adelaide

In Adelaide, research on community attitudes to water reuse has focused on recycled stormwater (as opposed to recycled wastewater, as in the Queensland and Western Australian contexts). The city of Adelaide sits upon several natural underground water aquifers which have historically been drawn from to supplement the water supply. Proponents of a recycled stormwater scheme have proposed an MAR process whereby treated stormwater is used to replenish aquifers which have become depleted. Human and environmental benefits are cited among the advantages of MAR schemes (which would allow water collection and storage during high rainfall periods, for subsequent retrieval and use during drier periods). However, there are several risks inherent in the MAR process. Among them is the potential for contamination of the aquifer due to poor quality of the water being introduced. Health risks are also of concern, given the potential for pathogens to remain in the water during the storage and retrieval processes.

Key findings

Despite the potential risks of MAR, community members were almost unanimously supportive of reusing stormwater (via the MAR process) for non-drinking purposes, and most people indicated they would also be willing to drink it (Mankad et al. 2013). Acceptance of stormwater reuse was explained by a range of policy-related variables. Hence, similar to the findings in relation to the south-east Queensland community acceptance of PRW, perceptions of effectiveness, fairness and trust in authorities emerged as important predictors of the acceptance of recycled stormwater for both potable and non-potable uses. Psychological variables also made a small contribution to the likelihood of stormwater acceptance, specifically descriptive norms (the degree to which a person believes that others in the community would use treated stormwater), moral norms (the sense of moral obligation towards reusing stormwater), favourable attitudes towards the benefits of stormwater, and concern for the long-term water security of the region.

Responses to and interactions with decentralised systems

Attitudes to decentralised systems: south-east Queensland

Historically, our urban centres have relied upon the assurances of centralised water supplies that are provided and maintained by governing authorities. As mentioned previously, the continuation of these supplies in south-east Queensland came under severe threat during the Millennium drought, when dam levels across Australia's eastern and southern states dwindled to record lows. In response to this threat, many homes were either constructed or retrofitted with technology to capture and reuse water. Rainwater tanks, ubiquitous in rural Australia, are now common in the urban landscape. Rainwater tanks typically capture and store rain that falls upon rooftops, for non-potable applications such as watering gardens and, if connected within homes, flushing toilets or washing clothes. Rainwater can also be used for drinking and cooking if extra filtering processes are included in the design. Another decentralised approach to water augmentation, greywater systems, diverts water for reuse, mostly from laundries and showers, that would otherwise be eliminated through the sewerage system.

Prior to the hard-felt impacts of the Millennium drought and the associated water restrictions, the majority of the wider community had been either unmotivated or actively resistant to installing decentralised infrastructure within their homes (Tjandraatmadja *et al.* 2008). However, decentralised water systems became far more widespread as residents became encouraged, either through the desire to be able to use water more freely in their homes, or by financial incentives in the form of government rebates for installing rainwater tanks. Moreover, within certain jurisdictions, new housing regulations stipulated the inclusion of mandatory water efficiency measures. These most commonly involved large rainwater tanks plumbed into the home to provide water for laundry, toilet flushing and outdoor purposes.

The social implications of decentralised systems were examined by conducting in-depth interviews with householders from three greenfield residential sites, which had incorporated decentralised technology into their design (Mankad *et al.* 2010). In households located within two of the sites, rainwater was used for potable applications as there was no centralised water supply. Households from neighbourhoods near the greenfield developments were also interviewed to gain their perspective on decentralised technology.

Key findings

Residents with decentralised systems generally reported that rainwater was superior to centrally supplied water for drinking, as it did not contain the chemicals which are routinely added to mains water. Other advantages included environmental benefits and not having to rely on the centralised water supply. Residents also believed that they would eventually save money by being able to provide their own water; however, it would take a long time (about 10 years) for them to recoup the costs of the technology and hence for the systems to 'pay for themselves'. There were also concerns raised, including maintenance issues and technology failure due to incorrect installation. In one household the residents were concerned that the technology problems meant that the water was no longer fit for drinking. Concerns were also expressed about the amount of electricity required to pump water from the tanks into the household.

Residents in neighbourhoods without decentralised systems generally conveyed acceptance of the technology, but provided strong arguments for why they did not have systems installed in their own homes or did not consider them to be an urgent necessity (Mankad and Tucker 2012). This response was in spite of the acknowledgement that water conservation was an important issue for society. The main barriers expressed were financial limitations, space constraints and the perceived complexity of the bureaucratic process of applying for government rebates (Mankad *et al.* 2010). Mistrust in the governance of decentralised water systems (and water systems more generally) was also cited by these householders.

Motivating rainwater tank maintenance

While owners of rainwater tanks have reaped the benefits of additional water availability, the ongoing costs and inconvenience of maintenance often receive much less attention. Regular maintenance of rainwater tanks (and the roof gutters which feed into them) is essential to avert potential health and environmental issues associated with water storage and reuse. Maintenance activities include clearing out leaves and debris from roof gutters, ensuring mosquito mesh is in good condition, testing water quality, maintaining pumps, and removing sediment from tanks. It has become evident, however, that not all rainwater tanks are subjected to regular maintenance, and we sought to determine why this might be.

Within south-east Queensland, rainwater tank owners can be broadly categorised into two groups: those who chose to retrofit their properties with a tank (often encouraged by government rebate schemes), and those who compulsorily acquired one upon building or purchasing a new property after January 2007. This second category of rainwater tank owners resulted from changes in the Queensland Development Code, stipulating that all newly constructed homes had to install water-saving devices to reduce reliance on the centralised water supply. This was most commonly achieved through the installation of rainwater tanks plumbed into the house for non-potable purposes (e.g. toilet flushing, clothes washing).

Key findings
We found that rainwater tank owners who chose to retrofit their property with a tank were more motivated to maintain their tank on a regular basis than those who became owners through regulatory requirements (mandated owners). Those who chose to install their tank displayed higher levels of self-determination, meaning that they engaged in the activities due to the personal satisfaction derived from doing so. In contrast, mandated tank owners were more likely to be amotivated to perform regular tank maintenance, meaning they did not feel a sense of control over the activities and perceived them to be meaningless. Consequently, retrofitted tank owners reported more frequent maintenance behaviours than mandated owners. They were also more likely to perceive the tank water to be a private resource that they could use at their own discretion, while mandated tank owners were more accepting of the notion that the government could regulate the tank water and potentially restrict its use.

These findings are concerning, and should serve as a resounding warning that recently developed residential sites may eventually harbour potentially harmful non-maintained rainwater tanks unless efforts are made to engage with and encourage tank owners to perform the necessary actions. As the memories of the Millennium drought recede and less community attention is given to rainwater tanks, there is a risk of these large receptacles being forgotten and falling into disrepair.

Water-sensitive urban design

Decentralised technology such as rainwater tanks and greywater systems fit within the sustainability framework of water-sensitive urban design (WSUD). Increasingly embraced within urban planning, the main objective of WSUD is to manage how water is used and discarded within communities. Successful applications of WSUD are able to minimise demands on the main potable supply and reduce the environmental impacts of stormwater and wastewater, particularly during times of water shortage. This is achieved through the implementation of water-efficient systems throughout communities including land and streetscaping measures, wetlands, and the use of alternative construction materials such as pervious pavements to allow for improved water drainage. While many WSUD features may be located in communal spaces, infrastructure such as rainwater tanks, greywater systems and water-efficient appliances can also be encouraged within individual homes that form part of WSUD communities.

Although there are numerous benefits of WSUD both in terms of water savings and reduced environmental impacts, the principles have not yet seen widespread adoption in mainstream urban development. In order to understand why, residents from six Adelaide communities with WSUD features participated in a series of interviews and focus groups (Leonard et al. 2014). The six communities varied in the scale of water-efficient technology that had been introduced within individual homes and public spaces in the broader neighbourhood. Four of the communities were new developments which had been purpose-built to incorporate a variety of eco-sensitive design features, while two were established neighbourhoods which had been retrofitted to incorporate WSUD features in communal areas.

Key findings

Among residents of WSUD communities, there was an overall sense of support and acceptance of WSUD, with improvements in community aesthetics and recreational amenity most often acknowledged as advantages. Residents with individual rainwater tanks and dual pipe systems were also appreciative of having more control over their water supply and believed this gave them licence to use water for outdoor purposes such as irrigation and washing cars. There was, however, far less general awareness of the environmental benefits of WSUD, such as improvement in water runoff quality and the potential for flood mitigation.

In each of the communities, technological failures were highlighted as one of the main barriers to more widespread WSUD adoption, with residents, councils and developers alike flummoxed by uncertainties and poorly functioning systems. These included water features that did not hold water, plants dying in bioretention installations, functional problems with non-potable water systems and inadequate maintenance of communal waterways. In communities with active residents' groups, there were high levels of knowledge and system monitoring. In contrast, there was a general lack of awareness in some neighbourhoods, due to poor community consultation. Hence, community education on WSUD systems and sustainable water use more generally was strongly advocated, with particular emphasis on engaging residents of retrofitted communities who are unlikely to be aware of the purpose of systems installed in their neighbourhood.

In conclusion, the community appears to be supportive of the general principle of WSUD, and adequate engagement and consultation should help to overcome any resistance associated with its application. Financial, technological and maintenance barriers are more likely to hinder its widespread adoption in urban development and renewal projects, than community objection.

Managing water demand: the other side of the equation

Water security does not rely solely on the provision of adequate water sources, but on the management and curtailment of demand. Put more simply – if urban residents use less water, then there will be less need to supply it. Indeed, the importance and expertise of social sciences is amplified by its potential for devising strategies to motivate people to reduce their water consumption.

Identifying drivers of household water demand: south-east Queensland

One of the first steps in managing the demand for water is to build a comprehensive picture of water use throughout the community. We need to know who is using the most water, how they are using it and what could possibly motivate them to use less. In order to do this, we conducted a large survey of households throughout south-east Queensland, asking owner/occupiers of free-standing dwellings how they used and thought about water (Fielding et al. 2012). We also collected information about who lived in the home and the types of water-using appliances that were present. To complete the picture, the water utility provided information about the amount of water used in each household for six months before and six months after the survey was completed.

Key findings

Not unexpectedly, the amount of water used in various households was influenced by a range of demographic, psychosocial, behavioural and infrastructure variables. Households in regions that had experienced fewer water restrictions, had more people living in the house and had higher income used more water. Although some water-efficient technologies were associated with lower water use (e.g. rainwater tanks plumbed into the house, low-flow taps), others were related to higher water use (e.g. water-efficient irrigation). This was most likely because the technology was associated with a particular water-related interest or hobby, e.g. a timed sprinkler system indicated householders regularly used water in their gardens.

Reducing demand: the potential of information-based strategies

After gaining an understanding of the psychosocial drivers of household water use, the next step was to motivate selected households to use less water in their homes (Fielding et al. 2013). This was achieved through an experimental design testing three types of information-based strategies (described in detail below). Information was provided on a series of four graphically designed postcards which were sent to participating households on a monthly basis. Highly detailed water usage of each home was observed for the duration of the intervention trial and for one year after the information had been provided (a total of 475 days). We also observed the water use of a fourth group of households (the control condition), who received no information. This last condition was included to discern whether any differences in water use could be attributed to general changes in the community, due to other variables such as weather patterns, time of year and so on. To facilitate the water use observation, all households had smart meters installed which provided daily total water consumption as well as the proportion of water used from each type of appliance (e.g. shower, toilet, washing machine etc.) (Beal et al. 2010). Households were selected for participation through their involvement in the large water use survey described above.

The three information strategies that were trialled were information only, descriptive norm, and water end-use feedback. Households assigned to the 'information only' condition received general water-saving tips about conserving water in the bathroom, laundry and kitchen, and fixing leaks. Households receiving the 'descriptive norm' information were provided with the same general water-saving tips plus information about what households like theirs (i.e. similar composition of adults and children) who were 'low water users' did in an effort to save water. The descriptive norm information had been derived from the large household survey which preceded the intervention trial. The rationale for testing this information strategy was based on prior research which found that information about what other people do is often more effective at achieving behaviour change than traditional moral-based appeals (Schultz et al. 2008). Finally, the 'water end-use feedback' households received both the general water-saving tips and tailored information detailing how their individual household used water. This information was provided in the form of a pie chart showing the proportion of water that was used in various appliances (e.g. the shower, toilet, taps etc.) throughout the house. The information was obtained through end use analysis of the data provided by individual smart meters (Beal et al. 2010).

Key findings

Relative to the control households which did not receive any information, there were significant reductions in household water use for all three types of information provided. This reduction peaked at about four months after the intervention (with savings of ~11.3 L per person per day), with water saving gradually dissipating over a 12-month period and returning to pre-intervention levels. Even after one year, however, the intervention households were still using less water than the control households, which had shown a marked increase in water use levels over that time. Surprisingly, there were no differences in the amount of water being used between households receiving the three different types of information, even though there had been an expectation that either the descriptive norm or the detailed end-use feedback information would be the most effective at altering behaviour.

We stress that the reductions in water use were quite remarkable, given the context of pre-existing low levels of water use throughout the community. That is, these households were already conservative with their water use following years of drought and water restrictions, yet managed to drop their consumption even further. This happened even while heavy rainfall deluged the region and widespread flooding occurred due to the dams overfilling. The drop in usage may have been partly due to the characteristics of the households participating in the research. It is reasonable to assume that these households were highly motivated to conserve water, given their willingness to respond to a survey on the issue then agree to participate in the intervention study. A similar campaign targeted to less water-conscious households may not have been as successful, or it may be that, among such recipients, the descriptive norm and detailed end-use information would prove more effective than the generic information strategy.

We can conclude that, in the absence of water scarcity and community restrictions, information-based strategies are effective for reducing household water consumption over the short to medium term. Over the longer term, additional strategies may be required to maintain water savings.

What have we learned?

Our research program has examined social issues that relate to both the supply of and demand for water within urban settings. More specifically, we have identified socio-demographic and psychological factors that influence community preferences for a range of potential water sources (e.g. recycled wastewater, stormwater) and acceptance of WSUD, explored communication strategies to facilitate understanding of alternative water management practices (e.g. recharging aquifers with urban stormwater), identified psychosocial drivers of household water use and other water-related behaviours (e.g. maintenance of rainwater tanks), and tested behavioural interventions that encourage more efficient use of water within the home.

Throughout the research described in this chapter, there were repeated themes which are worth noting. In particular, perceived fairness of decision processes and trust in authorities emerged as important determinants of community acceptance of various water technologies, including recycled wastewater. There was also a strong indication that the habits and culture surrounding the use of water had undergone a major disruption. This was because the research was conducted at a time when the country had been affected by the most prolonged and widespread drought in Australia's recent history, an experience which awakened a profound awareness of climate and sustainability issues, and catapulted the importance of water conservation into the forefront of the urban psyche. This wider context may have driven an overall acceptance of water-related technologies that previously may have been dismissed as unnecessary or too costly.

Our research demonstrated the importance of understanding not only how people think and behave in relation to water, but also their interconnectedness with social, technological and ecological systems. Although the theoretical basis of our research was underpinned by psychological paradigms which tend to focus on the role of the individual, we also explored ways in which the infrastructure and institutions that surround households have shaped patterns of consumption and the acceptability of water technology within the community.

Theoretical and practical implications

The projects we describe show how individual behaviour and attitudes are influenced by the broader cultural context, and that the relationships between people and water need to be interpreted within the social and technological systems in which they exist. That is, individuals do not behave within an isolated vacuum – their decisions and attitudes are driven by the culture in which they are embedded and are influenced by constructs such as fairness and trust.

Indeed, trust emerged as a major theme in our research, especially in relation to the provision of alternative water supplies. This is perhaps not surprising, as alternative water supplies are relatively new to Australians and there are inherent risks associated with the related technological processes. The literature is clear that trust is a major determinant of risk perceptions (Löfstedt and Cvetkovich 2008) and this is definitely echoed in our research. The literature also highlights that one way to develop trust is through using fair procedures that, for example, provide transparent information and give communities a voice so that the varying perspectives within society are reflected in the eventual decisions made by authorities (Tyler 2006). Hence, it makes theoretical sense that risk, trust and fairness are common themes in research that has focused on a resource that is essential to people's lives.

These findings have clear implications for authorities charged with managing water resources. Authorities need to work to develop trust in their organisation; the more customers and the general community trust the authority to responsibly manage water, the more likely it is that community members will embrace the introduction of new water management approaches rather than protest against them. Procedures that allow voice, consistency, accuracy, correctability, ethicality, representativeness and bias suppression are considered to be fair (Lind and Tyler 1988).

There is growing awareness of the need for these types of processes to engage communities and build trust; for example, the Western Australia Water Corporation's groundwater replenishment trial (Water Corporation 2014) provided ample information to the community and allowed two-way engagement through a visitor centre, a Facebook page, and a community portal that provided the opportunity to ask questions and engage in debate. Results of the water quality monitoring were provided through an easy-to-understand traffic light system to indicate water quality status (green = performance is acceptable and recharge is continuing, amber = an investigation has occurred and recharge is continuing, red = an investigation has occurred and recharge has stopped). Another example of a water organisation taking fair procedures seriously is the Metropolitan Water Directorate in New South Wales, which undertook a community engagement process to develop its Lower Hunter Water Plan (Metropolitan Water Directorate 2014). It conducted workshops to identify the values most important to the community about water planning, and had a transparent planning process that advised workshop participants on how their input would be used in decision-making.

Another theme that emerged in our research was the important contribution that can be made by an informed and motivated community of water users. When people have the necessary knowledge and motivation, they can help to respond effectively to drought situations, manage their onsite water infrastructure to avoid system failure and potential health risks, and make informed decisions about ideal water portfolios for their region. The challenge for organisations that manage water is how to increase levels of knowledge and awareness and increase motivation – especially in a world where people are already experiencing information overload and high levels of stress and pressure from modern life. Recent drought experiences have shown that Australians can respond well to public information campaigns, especially when environmental cues attune them to the issues (Walton and Hume 2011). A key concern when developing communication strategies is to consider the audience and give ample thought to how best to reach various segments of society. For example, developing communication strategies for regions that are marked by high levels of new immigrants may need to identify community leaders who can be a conduit for the communication process. Indeed, following the WSUD project, there appear to be benefits from a general strategy of tapping into a community's existing social capital by engaging with environmental and other community groups. We know that motivation can be increased if we meet people's need for autonomy, competence and relatedness, and these tenets should therefore be kept in mind when thinking about how to encourage community engagement with water-related issues (Deci and Ryan 2000). More research is needed to test how these principles can be put into practice to guide strategies that seek to engage community members with sustainable urban water management.

Conclusion

Our work has successfully integrated social science methods with technological approaches to solving the dilemma of water security. The research described in this chapter emphasises the central role of social and psychological considerations in sustainable urban water management. Water is a critical issue in Australia and around the world, and is likely to become more so in the future. We provide evidence that, when considering how to respond to current and future water challenges, it will be imperative not only to pursue technological solutions but also to consider how members of urban communities will interact with the technology and how they can become a force for positive change. Through applying and advancing theoretical concepts, this body of work has informed best practice in sustaining and protecting the vital substance which sustains our way of life.

References

Alexander KS, Price JC, Browne AL, Leviston Z, Bishop BJ, Nancarrow BE (2008) *Community Perceptions of Risk, Trust and Fairness in Relation to the Indirect Potable Use of Purified*

Recycled Water in South East Queensland: A Scoping Report. Technical Report No. 2. Urban Water Security Research Alliance.

Beal CD, Stewart RA, Huang T (2010) *South East Queensland Residential End Use Study: Baseline Results – Winter 2010.* Technical Report No. 31. Urban Water Security Research Alliance.

Cuddy SM, Maheepala S, Dandy G, Thyer MA, Hatton MacDonald D, McKay J, Leonard R, Bellette K, Arbon NS, Marchi A, Kandulu J, Wu W, Keremane G, Wu Z, Mirza F, Daly R, Kotz S, Thomas S (2014) *A Study into the Supply, Demand, Economic, Social and Institutional Aspects of Optimising Water Supply to Metropolitan Adelaide: Preliminary Research Findings. Summary Report from Project U2.2.* Technical Report Series No. 14/20. Goyder Institute for Water Research, Adelaide.

Deci EL, Ryan RM (2000) The 'what' and 'why' of goal pursuits: human needs and the self-determination of behavior. *Psychological Inquiry* **11**, 227–268. doi:10.1207/S15327965PLI1104_01

Dolnicar S, Hurlimann A (2009) Drinking water from alternative water sources: differences in beliefs, social norms and factors of perceived behavioural control across eight Australian locations. *Water Science and Technology* **60**, 1433–1444. doi:10.2166/wst.2009.325

Fielding K, Spinks A, Russell S, Mankad A (2012) *Water Demand Management Study: Baseline Survey of Household Water Use (Part B).* Technical Report No. 93. Urban Water Security Research Alliance.

Fielding K, Spinks A, Russell S, McCrea R, Gardner J, Stewart R (2013) An experimental test of voluntary strategies to promote urban water demand. *Journal of Environmental Management* **114**, 343–351. doi:10.1016/j.jenvman.2012.10.027

Inman D, Jeffrey P (2006) A review of residential water conservation tool performance and influences on implementation effectiveness. *Urban Water Journal* **3**, 127–143. doi:10.1080/15730620600961288

Leonard R, Walton A, Koth B, Green M, Spinks A, Myers B, Malkin S, Mankad A, Chacko P, Sharma A, Pezzaniti D (2014) *Community Acceptance of Water Sensitive Urban Design: Six Case Studies.* Technical Report Series No. 14/3. Goyder Institute for Water Research, Adelaide.

Leviston Z, Price J, Nicol S, Browne A, Nancarrow B (2009) *Management Strategies for Community Perceptions of Risk in the Reuse of Wastewater for Irrigation and Indirect Potable Supply.* CSIRO Water for a Healthy Country Flagship, Canberra.

Leviston Z, Browne AL, Greenhill M (2013) Domain-based perceptions of risk: a case study of lay and technical community attitudes toward managed aquifer recharge. *Journal of Applied Social Psychology* **43**, 1159–1176. doi:10.1111/jasp.12079

Lind EA, Tyler TR (1988) *The Social Psychology of Procedural Justice.* Plenum Press, New York.

Löfstedt RE, Cvetkovich G (2008) *Risk Management in Post-trust Societies.* Routledge, London.

Mankad A, Tucker D (2012) Alternative household water systems: perceptions of knowledge and trust among residents of south-east Queensland. *Ecopsychology* **4**, 296–306. doi:10.1089/eco.2012.0050

Mankad A, Tucker D, Tapsuwan S, Greenhill MP (2010) *Qualitative Exploration of Beliefs, Values and Knowledge Associated with Decentralised Water Supplies in South East Queensland Communities.* Technical Report No. 25. Urban Water Security Research Alliance.

Mankad A, Walton A, Leonard R (2013) *Public Attitudes towards Managed Aquifer Recharge and Stormwater Use in Adelaide.* Technical Report Series No. 13/10. Goyder Institute for Water Research, Adelaide.

Metropolitan Water Directorate (2014) *Planning for the Lower Hunter: Community Engagement.* <http://www.metrowater.nsw.gov.au/planning-lower-hunter/community-engagement>.

Price J, Fielding K, Leviston Z, Bishop B, Nicol S, Greenhill M, Tucker D (2010) *Community Acceptability of the Indirect Potable Use of Purified Recycled Water in South-East Queensland. Final Report of Monitoring Surveys.* Technical Report No. 19. Urban Water Security Research Alliance.

Price J, Fielding K, Leviston Z (2012) Supporters and opponents of potable recycled water: culture and cognition in the Toowoomba referendum. *Society and Natural Resources* **25**, 980–995. doi:10.1080/08941920.2012.656185

Schultz PW, Khazian A, Zaleski A (2008) Using normative social influence to promote conservation among hotel guests. *Social Influence* **3**, 4–23. doi:10.1080/15534510701755614

Shove E (2003) *Comfort, Cleanliness and Convenience: The Social Organization of Normality.* Berg, Oxford.

Tjandraatmadja G, Cook S, Sharma A, Diaper C, Grant A, Toifl M, Barron O, Burn S, Gregory A (2008) *ICON Water-sensitive Urban Developments.* CSIRO Technical Report. CSIRO National Research Flagships, Melbourne.

Tyler TR (2006) Psychological perspectives on legitimacy and legitimation. *Annual Review of Psychology* **57**, 375–400. doi:10.1146/annurev.psych.57.102904.190038

Walton A, Hume M (2011) Creating positive habits in water conservation: the case of Queensland Water Commission and the Target 140 campaign. *International Journal of Nonprofit and Voluntary Sector Marketing* **16**, 215–224. doi:10.1002/nvsm.421

Water Corporation (2014) *Groundwater replenishment* <http://www.watercorporation.com.au/water-supply-and-services/solutions-to-perths-water-supply/groundwater-replenishment>.

10

Making sense of Australians' responses to climate change: insights from a series of five national surveys

Iain Walker, Zoe Leviston, Rod McCrea, Jennifer Price and Murni Greenhill

How shall we understand opinions about climate change?

Climate change is one of the defining issues of modern times. Public and private opinions about climate change have shaped elections, mobilised social movements supporting and opposing action on climate change, and sparked changes in individuals' and communities' behaviours. Opinions about climate change matter.

In this chapter, we summarise some highlights from our five-year national research program on public understandings of climate change in Australia.

Psychology can contribute uniquely to an understanding of how human thoughts, feelings and actions contribute to environmental degradation generally, and especially to the formulation and implementation of effective interventions. Too many interventions at the policy and institutional level fail to deliver their desired effects because they fail to countenance the vagaries of humans. If we collectively are to understand how to mobilise effective behaviours to mitigate against climate change, and to adapt to inevitable climate change, then an adequate psychology is necessary (Clayton *et al.* 2015; Gifford 2014; Stern 1992; Swim *et al.* 2011). An understanding of climate change opinions is, in our view, a necessary part of any adequate psychology in this domain.

Overview

The CSIRO's Climate Adaptation Flagship acknowledged that human responses to climate change are a fundamental part of any sustainable solution to the challenges of climate change, and accordingly invested in a significant program of basic research aimed at better understanding human responses at individual, local, and institutional levels. Within this program of research, we conducted a series of five annual surveys of the Australian public's understanding of climate change, from attitudes, feelings, cognitions, and values, to engagement in climate-relevant behaviour, adaptation activities and support for adaptation policies. We summarise here some of the highlights of that work.

Our surveys were conducted online in July and August of each year from 2010 to 2014, inclusive. The sample size was ~5000 in each year and totalled 17 493 individuals across the five years, 4999 of whom completed two or more surveys; 269 of those completed all five. Respondents came from metropolitan, regional and rural Australia. Methodological details of the surveys are presented in each year's report (Leviston *et al.* 2013a, 2014; Leviston and Walker 2010, 2011) and in an end-of-project report (Leviston *et al.* 2015).

Our research shows how cognitive biases lead people to overestimate the prevalence of their own environmental views, the commonness of climate change denial in the broader community, and their personal contribution to the mitigation of environmental impacts. We find pro-environmental behaviours are not adequately explained by environmental attitudes; using longitudinal data, we show that behaviour predicts subsequent attitudes just as strongly as attitudes predict subsequent behaviours. Responses to climate change are strongly embedded in broader patterns of political ideology and worldview, and we show how attitudes to climate change can be analysed at the electoral-district level to predict swings in voting behaviour. Climate change responses are also strongly connected to broader worldviews, and we show how a reconceptualised model of myths of nature, based on cultural theory's framework of risk, explains a range of environmental behaviours.

Some key findings
What people think about climate change

A simple question appeared in all five surveys, asking respondents to indicate which of four different statements best described their thoughts about climate change. Response options were: I don't think that climate change is happening; I have no idea whether climate change is happening or not; I think that climate change is happening, but it's just a natural fluctuation in Earth's temperatures; I think that climate change is happening, and I think that humans are largely causing it. For ease, we will refer to these four response options as 'Not happening', 'Don't know', 'Happening but natural' and 'Happening and human-induced'.

Figure 10.1 shows the breakdown of responses to this question in each year. The proportions of respondents providing each response option appear to be relatively stable, with slight increases in the proportion of respondents who do not accept that climate change is happening and corresponding slight decreases in the proportion who accept climate

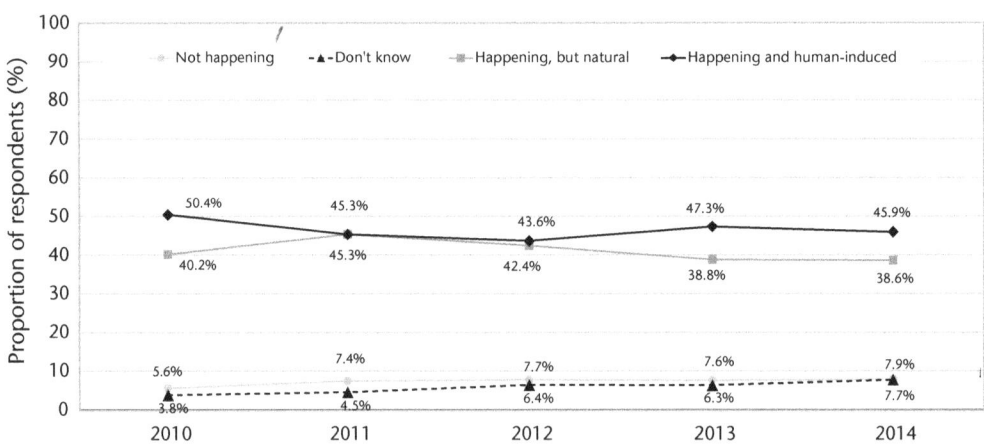

Fig. 10.1. Percentage breakdown of respondents' opinions about whether or not climate change is taking place from each annual survey (*n* = 5036, 5030, 5081, 5219 and 5163 in each year, respectively).

change is happening, regardless of cause. This is consistent with other poll data (e.g. Essential Reports 2016) and with our conclusion from a review of surveys done in Australia and overseas (Leviston et al. 2011): opinions about climate change have ebbed and flowed a little, but have not shifted radically in the last decade.

This fairly stable pattern of responses masks some intriguing and important inconsistencies, though. We look at two kinds of inconsistency: among the 269 respondents who participated in all five surveys, we can look at whether they change their responses to the same question from one survey to the next; and for all participants in any one survey (we will use the final survey for illustration), we can look at whether the responses to this question are logically consistent with similar questions about the same topic.

Using data from the 269 respondents who participated in all five surveys, Table 10.1 shows the proportion who offered the same answer in all five surveys. Not one of the respondents chose the *not happening* response in all five surveys, and almost none (0.4%) of the *don't know* respondents consistently indicated that they didn't know. Of the two dominant response categories, almost 22% of the *happening but natural* and almost 30% of the *happening and human-induced* respondents consistently provided that answer. Almost half (48.5%) of the 269 respondents changed their response at least once. Between the 2013 and 2014 surveys, 29% of respondents changed their response.

The inconsistency of responses across time suggests greater lability of climate change opinions than we suspect most researchers and commentators imagine. More important, though, is the direction of changes in opinion. Changes across time are not random. Table 10.2 shows the consistency in responses between the 2010 and the 2014 surveys (the most temporally distant responses). Entries in the cells on the leading diagonal represent respondents providing the same response in both surveys: 71% of respondents gave consistent responses in the two surveys. Those who migrated into the *human-induced* opinion group in 2014 came mostly from the *happening but natural* opinion group in 2010; none of the 11 people who responded in 2010 that they *don't know* if climate change is happening subsequently gave an opinion that climate change is *human-induced*, but three of the 12 people who in 2010 denied climate change is happening subsequently indicated that climate change is *human-induced*. Those who migrated away from the *human-induced* opinion group in 2010 were more likely to move into the *happening but natural* group in 2014 (30 of 129 people, 23%); two changed their response to *don't know*, and one indicated in 2014 that they now believed that climate change is *not happening*.

These patterns of opinion suggest that about two-thirds of the population have now developed a fairly stable opinion about climate change, divided evenly between those who accept that it is caused mostly by human activity and those who think it is a natural fluctuation in Earth's temperatures. The other one-third of respondents vacillate, but mostly between these two common opinion types. Of the *happening but natural* respondents in 2010 who had changed their opinion in the 2014 survey, 15 had shifted to the *human-*

Table 10.1. Percentage of respondents selecting the same statement about the causes of climate change in all five surveys (n = 269)

Response	Consistent across surveys (%)
Not happening	0
Don't know	0.4
Happening but natural	21.6
Happening and human-induced	29.4
All	51.5

Table 10.2. Cross-tabulation of respondents' climate change opinions from 2010 and 2014 surveys (n = 269). Pink-shaded cells represent those who responded with the human-induced opinion in 2010 but subsequently changed. Green-shaded boxes represent those who moved towards the human-induced opinion in 2014. Grey-shaded boxes along the diagonal represent those whose responses were consistent between 2010 and 2014

		2010 survey				
		Not happening	Don't know	Happening but natural	Happening and human-induced	2014 total
2014 survey	Not happening	1 (0.4%)	3 (1.1%)	9 (3.3%)	1 (0.4%)	14 5.2%
	Don't know	0 (0%)	3 (1.1%)	2 (0.7%)	2 (0.7%)	7 2.6%
	Natural	8 (3.0%)	5 (1.9%)	91 (33.8%)	30 (11.2%)	134 49.8%
	Human-induced	3 (1.1%)	0 (0%)	15 (5.6%)	96 (35.7%)	114 42.4%
	2010 total	4.5%	4.1%	43.5%	48.0%	100%

induced category and nine had shifted to thinking climate change was *not happening*. In contrast, of the *human-induced* respondents in 2010 who had changed their opinion by 2014, almost all (30) shifted to the *happening but natural* response, with two claiming they now *don't know* and just one denying climate change altogether. Thus, once people have developed a *human-induced* opinion about climate change, if they shift their opinion they are most likely to shift to the *happening but natural* view and are very unlikely to become climate change deniers. However, once people have developed an opinion that climate change is a natural process, a shift is likely either to a *human-induced* view or to a denial view, although they are slightly more likely to move to the former. It is as though those with a *human-induced* opinion have a latitude of acceptance (Sherif and Hovland 1961) that does not include denying climate change, but those with a *happening but natural* view have a broader latitude of acceptance that includes both a view that climate change is human-induced and a view that denies climate change altogether.

This has clear implications for those concerned with trying to communicate with the public about climate change and with mobilising support for changes aimed at reducing CO_2 emissions (Moser 2016). Producing broad change in public opinion need not be aimed at increasing acceptance of the role of human activity in driving climate change, at least not immediately or directly. That goal can be achieved by initially increasing acceptance that climate change is indeed a phenomenon, which may eventually be followed by increasing acceptance of the role of human activity.

So far in this section we have looked at the consistency and inconsistency in opinions about climate change from the same people across time. Now we move to look at consistency and inconsistency from the same people at the same time. For ease, we present results from only the 2014 survey, but similar findings were obtained in each of the five surveys.

Recall the results presented in Fig. 10.1, which show how people respond to the question 'Which of the following four statements best describes your thoughts about climate change?' A little earlier in the survey, respondents were presented with a slightly different question: 'Do you think that climate change is happening?' If people were logically consist-

Table 10.3. Cross-tabulation of respondents' climate change opinions with their responses to the question 'do you think that climate change is happening?' (2014 survey)

		\multicolumn{4}{c}{'Which of the following statements best describes your thoughts on climate change?'}	
		Not happening	Don't know	Happening but natural	Human-induced	Total
'Do you think that climate change is happening?'	Yes	36 (8.8%)	188 (47.5%)	1458 (73.2%)	2348 (99.1%)	4030 (78.1%)
	No	371 (91.2%)	208 (52.5%)	533 (26.8%)	21 (0.9%)	1133 (21.9%)

ent, the percentage answering *yes* to this question should be the same as the percentage who gave either the *human-induced* or *happening but natural* answer to the question summarised in Fig. 10.1. They are not: 78% answer *yes* to the question 'do you think that climate change is happening', while 84.5% answer either *human-induced* or *happening but natural* to a subsequent question. Indeed, when we cross-tabulate responses to these two questions, ~10% of all responses are inconsistent (see Table 10.3).

Another clear sign of people's inconsistency comes from looking at responses to a question asking people to indicate how much human activity contributes to climate change (see Fig. 10.2). Across all respondents, the average is that ~62% of climate change is seen to be the result of human activity. For those whose opinion is *happening and human-induced* the average is 79.2% (not 100%); for those whose opinion is *happening but natural* it is ~47% (not 0%); for those who deny climate change is happening at all it is ~35% (not 0%).

Similar sorts of discrepancies and contradictions arise when we look at how climate-change opinions are related to judgements about who is responsible for causing climate change. Figure 10.3 shows that even *happening but natural* people believe, on average, that big-polluting countries, multinational corporations, wealthy countries and federal government are at least 'partly responsible' for causing climate change. 'Individuals' and 'local government' are seen as less responsible, but the mean is still a fair distance above a score indicating 'not at all responsible'. These social agents probably do not fit into most people's categorisation of 'natural' causes of climate change.

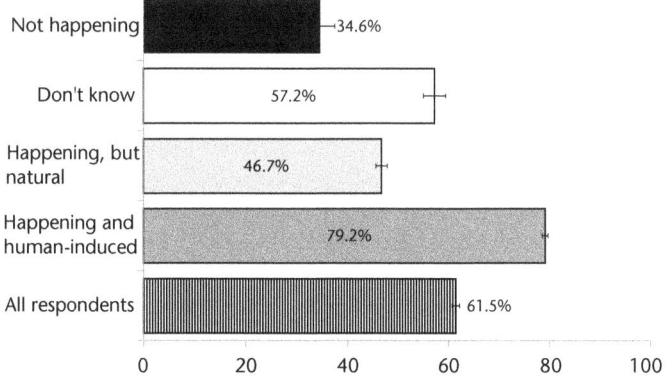

Fig. 10.2. Average estimated percentage contribution of human activity to climate change (2014 survey, *n* = 5163).

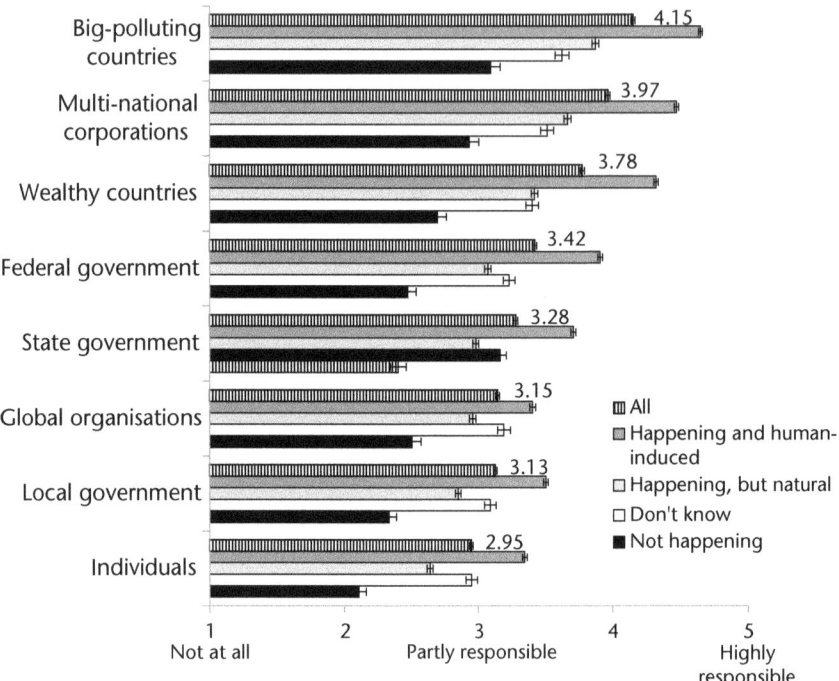

Fig. 10.3. Mean responses to the question 'Who is responsible for causing climate change?' for each of the four climate change opinion groups

Finally, Table 10.4 shows a cross-tabulation of opinions about climate change using different measures – our own opinion measure, and a measure that allows more combinations of human and natural processes to co-produce climate change (Reser et al. 2011). In a separate survey designed to test the response effects of question wording ($n = 897$), we used a split-sample design to compare within-respondent answers to these two different questions (Greenhill et al. 2014). Our results suggest that endorsing an opinion that sounds absolute and unequivocal ('climate change is not happening', 'climate change is happening but it's just a natural fluctuation in Earth's temperatures') does not necessarily indicate certainty of that opinion: prevarication, qualification and complication can underlie unequivocal opinions. How can this be explained, and what are the implications?

This apparent paradox of inconsistency in climate change opinions is difficult to reconcile with standard theories of attitudes (Eagly and Chaiken 1993), which are articulated at an intra-individual level of analysis and tend to assume an internal drive for consistency – there is too much variability across time and, especially, across questions, to support a view that people have coherent, consistent attitudes to climate change. An alternative viewpoint is that people have no coherent, consistent opinion about climate change, or indeed about anything, since expressed opinions are worked up discursively *in situ* (Potter and Wetherell 1987). That might help explain some of the variability, but there is too much consistency for this to be a wholly convincing position either. How, then, are we to understand the inconsistency?

We suggest that the meaning of the term 'climate change' is probably fluid, shifting according to question wording, what came earlier in the questionnaire, what respondents might think the researchers think or expect, and so on. We suspect people tend to assume

Table 10.4. Comparisons of responses to differently worded questions about climate change (n = 897)

Cause	Opinion group			
	Not happening (%)	Don't know (%)	Happening but natural (%)	Happening and human-induced (%)
Entirely natural	1.1	0.5	4.5	0.5
Mainly natural	1.1	0.7	10.1	0.5
Mixed	0.2	2.7	25.7	24.2
Mainly anthropogenic	0	0	0.6	16.7
Entirely anthropogenic	0	0	0	4.0
Not happening	2.9	0.3	0.7	0.1
Don't know	0.1	1.0	0.7	0.3
No opinion	0	0.3	0.3	0.2

that 'climate change' means 'anthropogenic climate change'. When we provide response options that qualify that default meaning (e.g. allowing them to indicate that climate change may be due to natural fluctuations in Earth's temperatures), people can cognitively move to a more nuanced view of what the term 'climate change' might mean. So, it is not necessarily psychologically inconsistent for people to respond differently to the question 'is climate change happening?' and then shortly after to the question 'which of the following four statements best represents your views about climate change?' and then later to questions about the degree of responsibility of different social agents for causing climate change. The responses can be 'inconsistent' because the object in question – 'climate change' – psychologically shifts.

For most people, and certainly for our many participants, 'climate change' is a social object. Most people do not develop beliefs and opinions about climate change through direct experience with the object 'climate change'. Rather, they develop beliefs and opinions through their experience with the social object 'climate change'. That object is socially constructed through media, through chatter in the café and among neighbours, and through political discourse. That object has come to be contested ideologically and thus has come to represent people's identities (Bliuc et al. 2015). People acquire a mental category 'climate change' through social interaction; they fill that mental category with content from social discourse; they valorise the category and its contents; they align themselves and identify, one way or another, with that category and its contents; they draw upon different meanings of that category in different contexts and for different purposes and functions.

Another reason why people respond to questions about climate change with apparent inconsistency is that their representation of climate change contains ambiguity, ambivalence, confusion and contradiction. There is no good reason to expect people to have representations of climate change that are internally consistent and coherent. Even climate scientists confess to ambivalence, simultaneously feeling hope and despair about what climate change means for humanity (http://www.isthishowyoufeel.com/).

A consequence for people who simultaneously hold inconsistent views is that it produces tension within them and their social systems, which in turn motivates change to reduce or adapt to that tension. Cognitive dissonance theory (Festinger 1957) shows how the discomfort from holding two psychologically discrepant cognitions motivates individuals to reduce the discomfort by changing either or both of those cognitions or by introducing a third. The inconsistency can be deeper than holding two specific cognitions, though. A study by Sapiains *et al.* (2015), partly based on the survey work presented in this chapter, shows that people experience a conflict between, on the one hand, their concern for the state of the environment, and, on the other, recognising that their consumer-oriented daily life is environmentally unsustainable. This tension leads people to dissociate, or separate, their concern for the environment from their unsustainable lifestyle practices, and to then accept their token behavioural gestures such as using reusable bags when grocery shopping as sufficient response.

Dissociation can be understood not just as an internal psychological process. It is a social process too, a form of denial. Literal denial of climate change is just one form of denial (Cohen 2001; Leviston and Walker 2012). It is possible to accept that climate change is happening, but interpret it in ways that distort its meaning or importance (interpretive denial), such as by disputing anthropogenic causation or by minimising the likely impacts on ecosystems and social systems. Another form of denial is implicatory denial. In this, the facts of climate change are not denied, nor are they interpreted to be something else. What is denied or minimised are the psychological, political and moral implications of the facts. We fail to accept responsibility for responding, we fail to act when the information says we should. The admixture of kinds of denial is apparent, we suggest, in the patterns of responses to questions about climate change.

What people think other people think about climate change

The fluidity of positions people can adopt in their opinions about climate change suggests that those opinions are serving different functions. Following cognitive dissonance theory, people's opinions are motivated, especially to resolve tensions arising from internally inconsistent and/or ambivalent views and from discrepancies between those views and their behaviours. A central tenet of cognitive dissonance theory (and many other theories) is that people are motivated to think well of themselves. Hence, people distort their perceptions of themselves, their behaviour and other people, to produce socially and/or morally palatable views of self. Evidence of some of these processes is revealed in the surveys when we ask people to position themselves relative to others, in terms of their opinions and their behaviours.

After answering the standard question that classified people into one of four opinion types, participants were asked to indicate what percentage of the Australian population held each of those four opinions. Figure 10.4 shows that, on average, respondents grossly overestimate the prevalence of rare opinions (*not happening* and *don't know*) and underestimate the prevalence of common opinions (*happening but natural* and *human-induced*). Figure 10.5 shows what happens when we break this pattern down according to participants' own opinion – a similar pattern of over- and underestimation is evident, but in addition people in each of the four opinion groups believe that their own opinion is the most common opinion in the community. This is particularly true for people in the rare opinion groups.

The findings are consistent in each of the five surveys; they are also consistent with a sizeable literature in social psychology on the false consensus (Ross *et al.* 1977; Mullen 1985; Nir 2011) and the pluralistic ignorance effects (Prentice and Miller 1996; Shamir and Shamir 1997; Leviston *et al.* 2013b). These effects occur for a mixture of cognitive, motivational and social reasons, such as being more likely to recall from memory attitude-consistent information, interacting with like-minded people, being exposed to media coverage

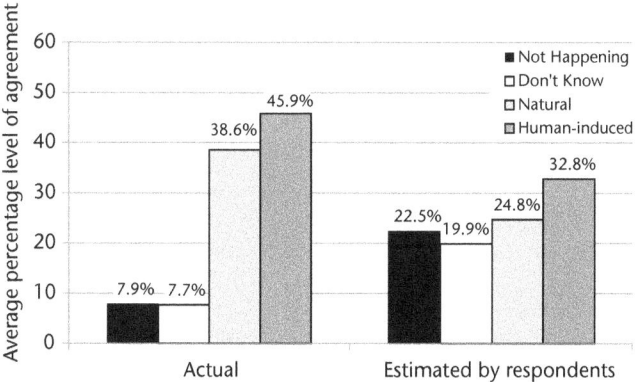

Fig. 10.4. Actual percentage of respondents professing each opinion type, compared to respondents' estimates of the prevalence of each opinion type in the community (2014 survey, n = 5163).

that is consistent with one's established opinions, and conforming to prevailing norms in one's membership and reference groups.

A similar self-serving bias is evident in people's responses to questions about pro-environmental behaviours, not just in their opinions about climate change. We asked participants to indicate whether they performed each of 21 different behaviours (did they recycle, did they turn off the lights when leaving a room, did they subscribe to Green Power electricity etc.), then asked how many of these 21 behaviours they thought they did compared to the general Australian population.

Some behaviours (e.g. switching off lights, trying to fix things rather than replace them) were commonly done but others (e.g. installing a greywater recycling system at home, contacting a government member about climate change) were rare. The average number of behaviours people reported fluctuated across years, from a high of 9.66 in the 2012 survey to a low of 8.96 in the 2014 survey. There was no clear trend over the five years. There was a consistent pattern, though, in how people thought they fared relative to others in the

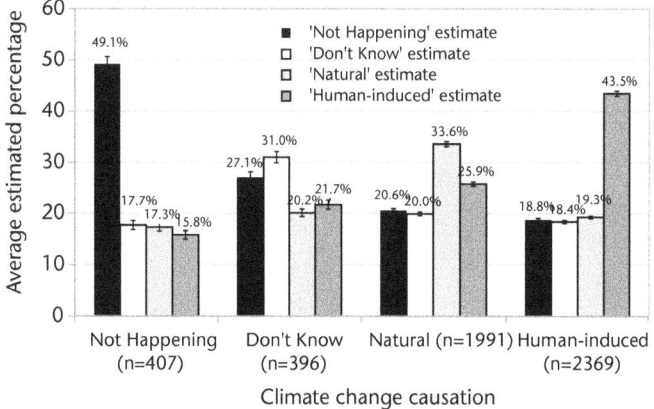

Fig. 10.5. Respondents' estimates of the community prevalence of each opinion type in the community, broken down by respondents' own opinion type (2014 survey, n = 5163).

Australian population. Few people (only 7% in the 2014 survey) thought they did fewer behaviours than others, a bit more than half (54.6%) thought they did about the same as others, and about a third (38.4%) thought they did more. It is statistically unlikely that only 7% of people do fewer behaviours than average, but it is common, in our data and across many other domains (Alicke *et al.* 2005; Dunning *et al.* 2003), for most people to think they do as well as, or better than, average.

Whether people think they do more behaviours than most people may have important implications for mobilising the individual, community and social behaviour changes necessary for effective mitigation of and adaptation to climate change. If people do relatively few behaviours but think they do more than average, they are likely to feel off the hook psychologically, free of the responsibility to do any more because they believe they are already doing more than their fair share (Tankard and Paluck 2016).

Links between opinions and climate-relevant behaviours, policies and worldviews

Some may argue that people's opinions about climate change and its causes matter little; it is their behaviours that count. But climate-relevant behaviours are related to opinions, and in non-obvious ways. In our survey data, we categorise behaviours into individual behaviours (the behaviours referred to above) and community behaviours (e.g. voting in a government election on the basis of an environmental issue, taking part in a conservation activity).

Individual behaviours are reported more frequently than community behaviours, but both are positively correlated with one another (0.46 in the 2014 survey). Both kinds of behaviour are more commonly reported by people with a *human-induced* opinion, and there are few differences in behaviour across the other three opinion groups. Individual and community behaviours are consistently and moderately related to a suite of climate-relevant attitudes, most strongly to how important and personally relevant a respondent perceives climate change to be.

It makes sense that opinions about climate change lead to behavioural outcomes. Our data suggest that that does happen. But so too do climate-relevant behaviours lead to opinions about climate change. In fact, behaviours predict subsequent opinions just as strongly as opinions predict subsequent behaviours. Figure 10.6 shows this, using data from respondents who participated in both the 2010 and 2011 surveys. (To conduct these cross-lagged correlations we constructed specific measures of climate attitude and behaviour, a bit different from the nominal opinion variable described above.)

Figure 10.6 shows that both the attitude measure and the behaviour measure were reasonably stable from one year to the next. In each year, attitudes and behaviour were associated with each other, more strongly in 2010 than in 2011. Importantly, and somewhat counterintuitively, the lagged correlations show that behaviours measured in 2010 predicted attitudes in 2011 just as strongly as attitudes measured in 2010 predicted behaviours in 2011. This perhaps appears odd, but it is entirely consistent with cognitive dissonance theory. People engage in behaviours, including climate-relevant behaviours, for many different reasons. Even people who deny that climate change is happening engage in climate-relevant behaviours. According to cognitive dissonance theory, a person who engages in a behaviour that counters a particular belief, attitude or opinion, will likely experience dissonance. People are motivated to reduce dissonance by making the discrepant elements more consistent, changing either element (the attitude or the behaviour) or introducing a third element. It is hard to deny the behaviour, once it has been done (though far from impossible! see Leviston and Walker 2012; Leitch 2011). An easier path to reduce disso-

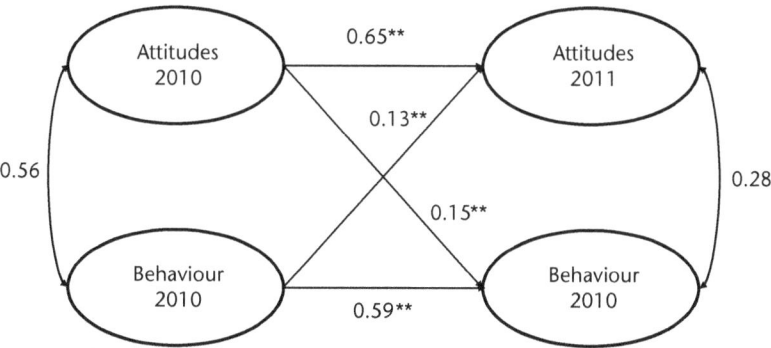

Fig. 10.6. Cross-lagged correlation between climate-relevant attitudes and behaviours, with data from respondents completing both the 2010 and the 2011 surveys.

nance is to change one's attitude. Alternatively, and alluded to above, a third element can be introduced – 'I believe that climate change is important and we should all do something about it', 'I don't do very much, but at least (I think) I do more than most other people'. In terms of the result in Fig. 10.6, people do act in accord with their attitudes, but they also bring their attitudes into accord with the behaviours.

We also considered two other important kinds of behaviour – voting behaviour and supporting government policy – and now discuss each of these.

Political behaviours, such as voting and mobilising for political action, may be among the more important behaviours we can engage in that have direct environmental outcomes. Climate change, its importance, and what, if anything, to do about it, have featured vividly in the Australian political landscape for a decade (Talberg et al. 2015; Smith et al. 2011). The left-leaning Australian Labor Party (ALP) won office in 2010, having played heavily on the need for action on climate change. The leader and new Prime Minister, Kevin Rudd, famously labelled climate change the greatest moral challenge of our time. The 2013 election saw power swing strongly back to the conservative Coalition, with leader Tony Abbott campaigning heavily to 'axe the carbon tax' (referring to the Clean Energy Legislative Package passed by then Prime Minister Julia Gillard, following Kevin Rudd's inability to get a previous package accepted by both houses of parliament).

Political parties' positions on climate change depend, in part, on the views of the electorate. In two studies, we have shown that the relationship between views about climate change in the electorate and voting outcomes are more complicated than political parties simply reflecting the views of their constituents. The views of the electorate partly depend on the positions of the parties they vote for.

In the first study (McCrea et al. 2015b) we modelled voting swings associated with changes in climate change beliefs at the individual level, and applied these results to the percentage of votes in each electorate at the 2010 Australian federal election. This enabled us to estimate how many electorates may be associated with a different outcome, given a hypothetical alteration in climate change beliefs across electorates. And rather than focusing only on scepticism about climate change, we constructed a new measure of attitude to climate change (different from the climate change opinions discussed above). This measure included avoidance, denial and pessimism, as well as scepticism. We found that climate change attitude and voting behaviour were associated such that, at the individual level, a hypothetical 10% change in climate change attitude (becoming more positive) was associated with a 2.6% swing away from voting for the conservative Coalition, and with a

2.0% sway towards the ALP and a 1.7% swing towards the Greens. Extrapolating this to the electoral level and modelling across the 2010 election results in each of the 150 electorates in the nation, this would result in a shift of 21 seats between the Coalition and the ALP after allocating Green Party preferences. The study suggests quite dramatically that apparently small changes in attitude, when linked to important behavioural outcomes like voting, can aggregate into large and significant social changes.

We extended this analysis in a second study (McCrea et al. 2015a), by aggregating data from the 2010 and 2011 surveys to the electorate level and combining them with 2010 voting outcome data at electorate level. The 2010 survey was conducted one month before the 2010 federal election, and contained a question about voting intention. In the 2011 survey (11 months after the 2010 election), we asked about actual 2010 voting behaviour and voting intention, although the next election was not scheduled until 2013. We again focused on climate change scepticism as the most important climate change belief for predicting voting behaviour (McCrea et al. 2011). Using cross-lagged modelling on both individual- and electorate-level data gave us a powerful test of whether voting behaviour influences scepticism (controlling for a host of other variables) or the other way round, or both. We found evidence that each influences the other, and that the influence of voting behaviour on scepticism is stronger than scepticism's influence on voting. This is somewhat counterintuitive, but consistent with the cognitive dissonance analysis mentioned earlier. It suggests that partisan political debates significantly influence how an electorate thinks about an issue.

We turn now from voting behaviour to policy support. In Australia, one of the most important legislative responses to climate change was the Clean Energy Legislative Package (CELP), commonly known as the carbon policy or carbon tax. The CELP introduced an emissions-trading scheme with a price for carbon, initially fixed but intended to eventually be floated. The CELP was enacted in July 2012 by the then federal ALP government. The CELP was strongly opposed by the conservative Coalition, which had previously supported an emissions-trading scheme but changed its position to support direct provision of incentives for abatement activities.

We conducted a survey of 616 Australian residents in November 2012, soon after the ALP government introduced the CELP (Dreyer and Walker 2013). More people found the CELP acceptable (43%) than unacceptable (36%; and a further 21% found it neither); 47% supported the policy and 53% opposed it, suggesting that almost all of those neutral on acceptability nonetheless opposed the policy. Policy support was predicted by how fair the policy was perceived to be, how effective it was perceived to be, and general free-market ideology.

A federal election was held in August 2013, and the CELP was a major election issue with the conservative Coalition campaigning to 'cut the carbon tax'. We decided to conduct another study to examine possible shifts in support for the CELP due to the election (Dreyer et al. 2015). We surveyed 772 participants a fortnight before the election, re-surveyed them a fortnight after the election (516 responded) and added 269 new participants to the post-election sample in order to assess any testing or measurement effects. Given how contentious the issue was, we were a little surprised that acceptance and support were both stable and relatively high across the election period. As before, acceptance was higher than support both pre- and post-election, and perceived effectiveness, perceived fairness and free-market ideology each predicted policy acceptance and support. A cross-lagged correlation analysis indicated that acceptance is a precondition of support.

The results discussed so far show clearly that climate change opinions influence, and are influenced by, climate-relevant behaviours. They also influence support for government policy to take action to reduce greenhouse gas emissions. Our surveys also show that climate change opinions are linked to broader worldviews; we now turn to briefly consider these links.

The concept of worldviews comes from cultural theory (Douglas 1978, 1985; Douglas and Wildavsky 1982), which provides a powerful framework connecting individual beliefs and values about society and the environment with conflicting social discourses about risk, and with policy preferences and behavioural choices.

Cultural theory asserts that competing views of what constitutes a social risk (e.g. nuclear power, secularism, unfettered technological development, environmental exploitation and damage, increasing inequality) are moral judgements about social order, not probabilistic assessments of the likelihood and severity of negative events. Patterns of shared beliefs and values are referred to as cultural biases (Wildavsky 1987), which in turn define four worldviews (hierarchical, egalitarian, individualistic and fatalistic).

Recently, Price *et al.* (2014) extended the cultural theory framework from its focus on patterns of beliefs about social relations to understand orientations to environmental risks. The study showed that environmental biases form two negatively correlated factors, which we describe as 'environment as elastic' and 'environment as ductile'. The former represents an individualised perspective and combines egalitarian and hierarchical views. It sees the environment as unaffected and uncontrollable by humans, as resilient and able to absorb the insults of human activity. There is no need for individual or collective actions to preserve or protect the environment. In the view of the environment as ductile, human activity alters the environment, which is unable to recover if it is damaged too much, and restraints on individual and human behaviours are justified in order to protect the fragile environment.

Beliefs that the environment is elastic or ductile are strongly related to opinions about climate change (see Fig. 10.7). The *not happening* group was neutral on both environmental positions. No opinion group endorsed the 'environment as elastic' view, but the *human-induced* group strongly disagreed with it. Both opinion groups that accepted climate change endorsed the 'environment as ductile' view, especially the *human-induced* group. When we correlate *ductile* and *elastic* with measures of engagement in individual behaviour, engagement in community behaviour and support for adaptation policy initiatives, *ductile* is strongly and positively correlated with all three behaviours, whereas *elastic* is negatively and more weakly correlated with community behaviours and policy support and only slightly and negatively correlated with individual behaviours. These associations with individual and community behaviours and with adaptation policy support remain even after controlling for a range of other attitudinal and emotional variables relevant to climate change.

Conclusions and implications

Opinions about climate change are important. They influence behaviours from engaging in individual-level mundane domestic behaviours through to voting, lending support to government policy, and engaging in collective action. Engaging in those behaviours in turn influences the opinions people hold and express. People do not adhere to a single unwavering opinion about climate change. People's opinions about climate change fall within a broad latitude of acceptance, and shift over time and situation. The orderly ways they shift have important implications for science communication and the mobilisation of support for action to mitigate and adapt to climate change. The constellation of each person's opinions about climate change is full of contradiction and uncertainty. People's opinions are patterned by broader ideologies and worldviews – they are not developed afresh, disconnected from how people think about other aspects of their social and environmental

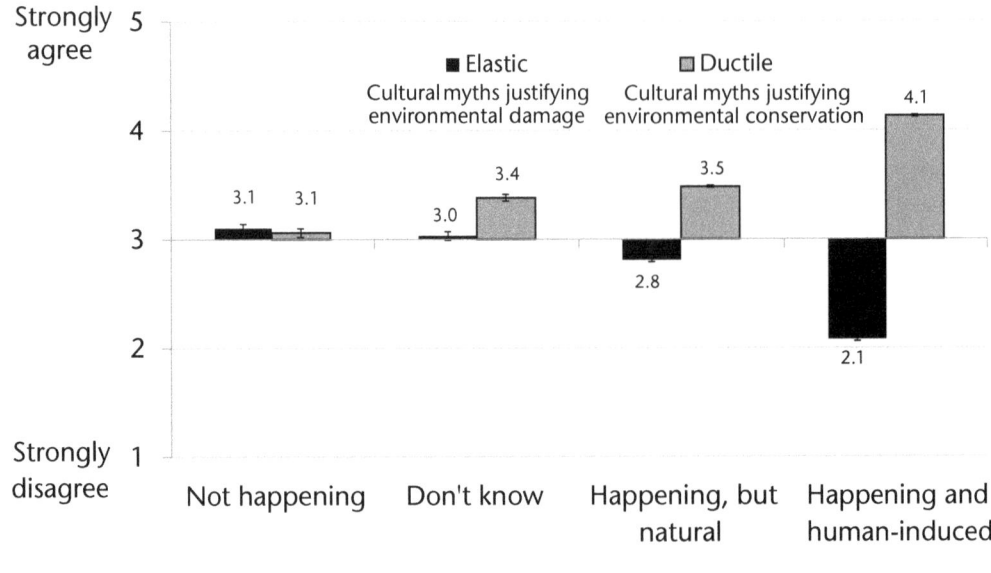

Fig. 10.7. Endorsement of *ductile* and *elastic* views, by climate change opinion (2014 survey, n = 5163).

worlds. Despite the importance of climate change opinions, people are remarkably poor in their knowledge of everyone else's opinion about climate change.

So, how are we to understand climate change opinions? They are best understood as the expression of one aspect of how people broadly and tacitly understand their relationship with the social and environmental worlds around them. They are psychological systems in tension, assimilating and accommodating new, sometimes difficult and challenging, information. They are functional for the people who express them, they serve as markers of social identity, and as systems they have inertia and a drive to persist – all of which makes them difficult, but not impossible, to change.

References

Alicke MD, Dunning DA, Kruger JI (2005) *The Self in Social Judgment*. Psychology Press, New York.

Bliuc A, McGarty C, Thomas E, Lala G, Berndsen M, Misajon R (2015) Public division about climate change rooted in conflicting socio-political identities. *Nature Climate Change* 5, 226–229. doi:10.1038/nclimate2507

Clayton S, Devine-Wright P, Stern PC, Whitmarsh L, Carrico A, Steg L, Swim J, Bonnes M (2015) Psychological research and global climate change. *Nature Climate Change* 5, 640–646. doi:10.1038/nclimate2622

Cohen S (2001) *States of Denial: Knowing about Atrocities and Suffering*. Polity Press, New York.

Douglas M (1978) *Cultural Bias*. Royal Anthropological Institute of Great Britain and Ireland, London.

Douglas M (1985) *Risk Acceptability According to the Social Sciences*. Russell Sage Foundation, New York.

Douglas M, Wildavsky A (1982) *Risk and Culture: An Essay on the Selection of Technical and Environmental Dangers.* University of California Press, Berkeley, CA.

Dreyer S, Walker I (2013) Acceptance and support of the Australian carbon policy. *Social Justice Research* **26**, 343–362. doi:10.1007/s11211-013-0191-1

Dreyer SJ, Walker I, McCoy SK, Teisl MF (2015) Australian acceptance and support for carbon pricing before and after the federal election. *Nature Climate Change* **5**, 1064–1067. doi:10.1038/nclimate2756

Dunning D, Johnson K, Ehrlinger J, Kruger J (2003) Why people fail to recognize their own incompetence. *Current Directions in Psychological Science* **12**, 83–87. doi:10.1111/1467-8721.01235

Eagly AH, Chaiken S (1993) *The Psychology of Attitudes.* Harcourt Brace Jovanovich, Fort Worth, TX.

Essential Reports (2016) Climate change. <http://www.essentialvision.com.au/climate-change-13>.

Festinger L (1957) *A Theory of Cognitive Dissonance.* Stanford University Press, Stanford, CA.

Gifford R (2014) Environmental psychology matters. *Annual Review of Psychology* **65**, 541–579. doi:10.1146/annurev-psych-010213-115048

Greenhill M, Leviston Z, Leonard R, Walker I (2014) Assessing climate change beliefs: response effects of question wording and response alternatives. *Public Understanding of Science (Bristol, England)* **23**, 947–965. doi:10.1177/0963662513480117

Leitch A (2011) Understanding emotional responses to climate change. *Ecos* 4 May.

Leviston Z, Walker I (2010) *Baseline Survey of Australian Attitudes to Climate Change: Preliminary Report.* National Research Flagships Climate Adaptation, CSIRO Ecosystem Sciences, Perth.

Leviston Z, Walker I (2011) *Second Annual Survey of Australian Attitudes to Climate Change: Interim Report.* National Research Flagships Climate Adaptation, CSIRO Ecosystem Sciences, Perth.

Leviston Z, Walker I (2012) Beliefs and denials about climate change: an Australian perspective. *Ecopsychology* **4**, 277–285. doi:10.1089/eco.2012.0051

Leviston Z, Leitch A, Greenhill M, Leonard R, Walker I (2011) *Australians' Views of Climate Change.* CSIRO, Canberra.

Leviston Z, Walker I, Malkin S (2013a) *Third Annual Survey of Australian Attitudes to Climate Change: Interim Report.* CSIRO, Perth.

Leviston Z, Walker I, Morwinski S (2013b) Your opinion on climate change might not be as common as you think. *Nature Climate Change* **3**(4), 334–337. doi:10.1038/nclimate1743

Leviston Z, Price J, Malkin S, McCrea R (2014) *Fourth Annual Survey of Australian Attitudes to Climate Change: Interim Teport.* CSIRO, Perth.

Leviston Z, Greenhill M, Walker I (2015) *Australian Attitudes to Climate Change: 2010–2014.* CSIRO, Canberra.

McCrea R, Leviston Z, Shyy T-K, Walker I (2011) *Scepticism Counts: Impacts of Climate Change Beliefs on Primary Votes for Political Parties at the 2010 Australian Federal Election.* Poster presented at the *Behavior, Energy & Climate Change Conference.* 30 November –2 December 2011, Washington DC.

McCrea R, Leviston R, Walker I (2015a) Climate change skepticism and voting behaviour: what causes what? *Environment and Behavior.* doi:10.1177/0013916515599571

McCrea R, Leviston Z, Walker I, Shyy T-K (2015b) Climate change beliefs count: relationships with voting outcomes at the 2010 Australian federal election. *Journal of Social and Political Psychology* **3**, 124–141. doi:10.5964/jspp.v3i1.376

Moser SC (2016) Reflections on climate change communication research and practice in the second decade of the 21st century: what more is there to say? *WIREs Climate Change* 7, 345–369. doi:10.1002/wcc.403.

Mullen B (1985) The false consensus effect: a meta-analysis of 115 hypothesis tests. *Journal of Experimental Social Psychology* 21, 262–283. doi:10.1016/0022-1031(85)90020-4

Nir L (2011) Motivated reasoning and public opinion perception. *Public Opinion Quarterly* 75, 504–532. doi:10.1093/poq/nfq076

Potter J, Wetherell M (1987) *Discourse and Social Psychology: Beyond Attitudes and Behaviour.* Sage, London.

Prentice D, Miller D (1996) Pluralistic ignorance and the perpetuation of social norms by unwitting actors. *Advances in Experimental Social Psychology* 28, 161–209. doi:10.1016/S0065-2601(08)60238-5

Price J, Walker I, Boschetti F (2014) Measuring cultural values and beliefs about environment to identify their role in climate change responses. *Journal of Environmental Psychology* 37, 8–20. doi:10.1016/j.jenvp.2013.10.001

Reser JP, Pidgeon N, Spence A, Bradley G, Glendon IA, Ellul M (2011) *Public Risk Perceptions, Understandings, and Responses to Climate Change in Australia and Great Britain: Interim Report.* Griffith University, Climate Change Response Program, Queensland/Understanding Risk Centre, Cardiff University, Cardiff.

Ross L, Greene D, House P (1977) The false consensus phenomenon: an attributional bias in self-perception and social perception processes. *Journal of Experimental Social Psychology* 13, 279–301. doi:10.1016/0022-1031(77)90049-X

Sapiains R, Beeton RJS, Walker I (2015) The dissociative experience: mediating the tension between people's awareness of environmental problems and their inadequate behavioral responses. *Ecopsychology* 7, 38–47. doi:10.1089/eco.2014.0048

Shamir J, Shamir M (1997) Pluralistic ignorance across issues and over time: information cues and biases. *Public Opinion Quarterly* 61, 227–260. doi:10.1086/297794

Sherif M, Hovland CI (1961) *Social Judgment: Assimilation and Contrast Effects in Communication and Attitude Change.* Yale University Press, New Haven, CT.

Smith TF, Thomsen DC, Keys N (2011) The Australian experience. In *Climate Change Adaptation in Developed Nations: From Theory to Practice.* (Eds JD Ford, L Berrang-Ford) pp. 69–84. Springer Science, Berlin.

Stern PC (1992) Psychological dimensions of global environmental change. *Annual Review of Psychology* 43, 269–302. doi:10.1146/annurev.ps.43.020192.001413

Swim JK, Stern PC, Doherty TJ, Clayton S, Reser JP, Weber EU, Gifford R, Howard GS (2011) Psychology's contributions to understanding and addressing global climate change. *American Psychologist* 66, 241–250. doi:10.1037/a0023220

Talberg A, Hui S, Loynes S (2015) *Australian Climate Change Policy: A Chronology.* Parliamentary Library Research Paper Series 2015–16. Australian Parliamentary Library, Canberra.

Tankard ME, Paluck EL (2016) Norm perception as a vehicle for social change. *Social Issues and Policy Review* 10, 181–211. doi:10.1111/sipr.12022

Wildavsky A (1987) Choosing preferences by constructing institutions: a cultural theory of preference formation. *American Political Science Review* 81, 3–22. doi:10.2307/1960776

11

Innovation, sustainability and the promise of inclusion

Lucy Carter

Sustainability science approaches demand new ways of working. Methods and tools which firmly place knowledge into use, build bridges across multiple boundaries and foster diversity of people and institutions have been strategies previously advanced by sustainability science scholars (Jerneck et al. 2011; Kates et al. 2001). Much of the literature to date has offered perspectives on conceptualising sustainability science and defining research approaches and roles (Wittmayer and Schäpke 2014). Less commonly shared have been the lessons and challenges presented by *doing* science in this way (Lang et al. 2012). In many ways, sustainability science is a reorganisation of traditional approaches to science. It demands new skills, new tools and a willingness to experiment. Lessons from scientists who have attempted to implement methodologies considered central to sustainability science are yet to emerge –this chapter contributes to this somewhat neglected area of work. Specifically, it offers insights into the individual, operational and institutional implications of pursuing a process-oriented approach to sustainability challenges.

The chapter is a personal reflection on the experiences and lessons from an ambitious project which attempted to create and use knowledge differently in order to address a sustainability challenge – food systems innovation. The knowledge generated has influenced Australian international development policy in the quest for more effective ways of 'doing development'. As became apparent in the course of the project, doing science differently can be very challenging, for a multitude of reasons. This chapter shares some of those challenges, in the spirit of informing future planning and practice.

A distinct lens has been chosen for this analysis. Gender considerations, as they apply specifically to sustainable development and more broadly to implementation and delivery science, have been selected to interrogate a process-oriented approach to sustainability science. The decision to apply a gender lens is explained in more detail below; it largely relates to the values, visions and goals implicit in sustainability science approaches. Goals and values which relate to integration (of science into policy and practice), diversity (of people, institutions and knowledge), equality and inclusion are intimately linked to research and practice aiming to effect sustainable change. Gender diversity and equality are central to this vision.

Based on experiences from a three-year project seeking to catalyse learning and innovation in food systems research and development, this chapter discusses the implications for individuals and organisations seeking to apply novel methods for bringing about institutional change. It identifies unanticipated costs (and benefits) of pursuing such approaches and considers the implications for men and women applying them in research organisations. Using themes from innovation science and gender mainstreaming, it argues that the principles and approaches central to sustainability science demand and benefit from specific institutional, organisational and individual conditions. Many of these require shifts in organisational culture, systems and structures before they enable the conditions necessary for impact.

Inclusion, integration and innovation: the emergence of implementation science

Global responses to meeting food and nutrition security have prescribed the use of inclusive, innovative approaches to the development process (Brooks and Loevinsohn 2011). The Global Alliance for Improved Nutrition (GAIN) and Leveraging Agriculture for Nutrition in South Asia (LANSA) are two examples of international initiatives which follow an innovation and inclusion agenda for improving food and nutrition security. The intent to redress past inadequacies partly drives this agenda, as does the need to reach scalable, sustainable impact using finite and ever-diminishing aid resources. Achieving institutional change through processes of learning and action is a key driver for applying innovation approaches to development.

Innovation systems approaches to agricultural research for development (R4D) offer a framework to support actor-driven and collaborative processes for learning and action (Foran et al. 2014). Tools and methods which facilitate the co-creation of knowledge and deliver solution-oriented outcomes are central to both defining and resolving sustainability challenges in development. In many ways, the science of innovation – translating and embedding research into social and economic activities – is the new frontier for R4D.

Efforts which stimulate cooperation across sectors and scales to address emerging global problems have also ascended from other distinct yet complementary areas of science. The One Health movement shares a common philosophical position with sustainability science (see Gibbs 2014 for a brief history of the One Health concept). Although less sharply articulated in the case of One Health, both approaches aim to improve societal understanding of the complex interactions between humans, animals and the environment, in an effort to address threats to sustainability. For One Health, the challenge is to curb local, regional and global infectious disease outbreaks. While the approaches differ in their theoretical maturity and the methodologies employed, both seek to redefine how knowledge is created and used.

Implementation research is also proving to be the new frontier for clinical and health sciences more broadly. Efforts which improve the effectiveness of health service delivery through the adoption of evidence-based practice have attracted significant investment in recent years and continue to attract multisectoral cooperation globally. With comparable goals and similarly complicated contexts, there perhaps lies opportunity for cross-learning in bridging the gap between problem-solving and innovation.

The tools most suitable for triggering innovation are those oriented to fostering collaboration and cooperation among diverse actors. For innovation in food systems, brokering – proactively facilitating a process towards a common goal – either for knowledge integration or for building effective partnerships, is a useful device for triggering institutional change (Klerkx and Gildemacher 2012). Similarly, transdisciplinary approaches for identifying adaptation pathways to climate change rely on inclusive, participatory processes to achieve impact.

Meeting sustainability goals using gender mainstreaming to achieve gender equality

In many ways, inclusiveness (of perspectives, roles and institutions) is a central theme in the post-2015 sustainable development agenda. For example, understanding and valuing the range of gender perspectives on sustainability challenges is key to realising more effective development outcomes. Njenga and colleagues' (2011) definition of gender best communicates the nuances that are intrinsic to gender descriptions but often missing in interpretations: 'Gender refers to the sociocultural constructs of roles, responsibilities, characteristics, attitudes and beliefs towards men and women. These relationships are learned, change over time and vary widely between cultures' (Njenga *et al.* 2011, p. 380).

Realising gender equality is a key goal in R4D, and mainstreaming gender across the programming and policy spectrums is recognised globally as a critical approach for achieving impact (Njenga *et al.* 2011; UN Entity for Gender Equality and the Empowerment of Women 2014). Failure to include both women's and men's perspectives in the planning, implementation and evaluation of interventions is at best potentially ineffective, and at worst detrimental to beneficiaries and other stakeholders of development interventions. For instance, striving to improve nutrition security at the household level in Timor-Leste will be less successful if men are not equal partners in any proposed intervention.

Despite decades of global effort, there has been mixed success in realising gender equality using mainstreaming approaches. Many commentators now argue that approaches to date have fallen short of transformational change. At the policy and programming levels, a continued overreliance on devising stand-alone gender policies to address inequality has not delivered anticipated change. Similarly, the provision of (mostly) context-neutral gender training to staff as the preferred strategy for institutionalising gender concerns in R4D practice has failed to achieve impact at scale (Porter and Sweetman 2005).

These approaches fall short as mechanisms for triggering institutional change for several reasons: they often fail to include men's perspectives, roles and responsibilities, and they tend to perpetuate the misconception that women are a homogenous group. Relationships between gender, ethnicity and age are often excluded in generic approaches to gender mainstreaming and gender capacity-building. Most disturbingly, such all-purpose approaches often neglect to take account of power in social relationships and its influence on decision-making capacity, autonomy and empowerment.

Since the historic *Beijing Platform for Action*, a set of UN resolutions that emerged from the 1995 World Conference on Women, proponents of gender mainstreaming have consistently contended that gender considerations must first be embedded at the organisational and operational levels before such strategies can successfully be implemented outwardly. In other words, gender mainstreaming approaches must be integrated into the policies, systems and procedures – the deep structures[1] – of implementing organisations before they can successfully achieve their goals using policy and programming channels. Accordingly, achieving gender equality depends on organisational maturity in tackling

---

1 A phrase coined by Rao and Kelleher (2005) to describe the collection of deeply embedded cultural values, behaviours, attitudes and ways of working that underlie decision-making and action within organisations.

diversity and inclusion in the workplace. This is a mammoth task for development donors and organisations seeking to improve development outcomes.

Gender mainstreaming strategies manifest differently across organisations and sectors. In principle at least, the realisation that gender equality is unlikely to be reached solely by using measures which create gender parity in recruitment and promotion is accepted by most large Australian organisations. The business sector has been faster than most to realise the value of achieving diversity and inclusion in the workplace (Sanders *et al.* 2011). For research organisations, the journey has been more difficult.

Striving for gender parity is, however, a step in the right direction. Globally, gender gaps in science and technology persist despite women's comparatively strong publishing presence in the social sciences (Castillo 2015). Structural barriers partly explain this gap, as do cultural obstacles – the stereotypes, biases and values which perpetuate prejudice. Gender differences in personality, leadership styles, cooperation and communication styles also play a role in the quality of experiences for both women and men in research organisations.

A long history of measuring science impact using bibliometric and citation data has been shown to perpetuate gender gaps (Castillo 2015). Women are systematically less cited than men, probably due to a combination of factors including a tendency for women to cite themselves less often than men, and men tending to cite male authors more often than female authors (Castillo 2015). For international R4D, where impact is best evaluated using additional measures of success such as the quality of partnerships created, the extent of capacity built or the depth of community engagement, the burden of proof is far greater.

For sustainability researchers, the tools and processes available to trigger and facilitate innovation demand supportive organisational structures and cultures. For organisations in pursuit of innovation as an impact pathway, mechanisms which motivate and reward efforts to innovate are more likely to generate impact. Yet despite the emergence of new methods to evaluate process-oriented impact, widespread adoption has been slow.

Innovation systems approaches, and participatory methods more broadly, which facilitate knowledge integration such as brokering, all require researchers to organise and cooperate differently. Researchers need to successfully engage with a range of stakeholders who are more diverse than ever. This has implications for how men and women experience and succeed in the workplace. Incentives and instruments to stimulate innovation must meet the needs of all scientists. The following section reflects on how these themes manifested in a multi-stakeholder partnership which sought to inspire food systems innovation both domestically and abroad.

The Food Systems Innovation initiative: a test in science working differently

The Food Systems Innovation (FSI) project attempted to trigger innovation at the institutional level by strengthening the evidence base on what is working in development and what is not, by using and applying analysis and by delivering capacity-building and learning activities. While triggering social learning using action research methods might have described our approach, FSI is essentially about 'doing development better'. For FSI, the knowledge generated must be usable, meaningful and inclusive of as many actors as feasible. The project ran for three years, ending in December 2015. While it is still early to definitively report on project successes, some general observations and reflections might be useful for practitioners attempting similar endeavours.

In partnership with the Department of Foreign Affairs and Trade (DFAT), the Australian Centre for International Agricultural Research (ACIAR) and the Australian International Food Security Research Centre (AIFSRC), FSI sought to bring together Australian and international expertise to improve the impact of agriculture and food security programs in the Indo-Pacific region. Over the three years, project partners and other stakeholders worked together to share ideas for improving development programming and policy decision-making in food systems research and practice. The project underwent considerable modification to design and governance models at least annually, partly in response to changing political and organisational conditions. Adaptive management has always been an important element of FSI.

For the last two years, project activities were aligned to key international development themes. These included managing for impact (theories of change and impact pathways); markets and partnerships (pro-poor markets and inclusive business models) and agriculture links (multisectoral cooperation to improve development outcomes). This author led the third stream of work – strengthening the pathways between agriculture and improved nutrition outcomes – a topic largely neglected until recently.

Nutrition-sensitive agriculture (NSA) approaches seek to enhance food and nutrition security with a view to improving nutrition and health outcomes. This approach reflects the continued prevalence of malnutrition in low- and middle-income countries despite increases in income and food production. Malnutrition in all its forms has cognitive, reproductive, economic and health impacts across the lifespan. Malnutrition threatens livelihoods and places significant health and economic burdens on people, communities and countries. NSA has been described as 'agriculture with a nutrition lens' –policy and programming are led by the agriculture sector in collaboration with other sectors, to reduce poverty and food insecurity and improve livelihoods. While NSA is a relatively new field of research and practice in international development, it shares many similarities with other emerging topics in agricultural development.

It is a mode of development practice and policy that sits at the interface of several research and practice traditions and stakeholder interests. NSA straddles the science–policy interface and demands knowledge across a range of sectors including agriculture, nutrition, food systems and health. While there is international consensus on investing in NSA-related activities, broad experience in applying and evaluating NSA approaches is yet to emerge. The challenges for NSA, not unlike other approaches seeking to improve development outcomes, is to bridge the knowledge gaps, navigate the multiple boundaries and more rigorously evaluate impact. In doing so, multisectoral cooperation is made easier, tensions are reduced and pathways for action become clearer.

NSA offers an ideal setting in which to practise sustainability science approaches. It demands collaboration between sectors and stakeholders, integration of multiple knowledge types, and process facilitation in order to stimulate innovation. Collaboration, transdisciplinarity and process facilitation play critical roles in bridging these interfaces, and brokering assumes importance.

To be fair, the FSI initiative never held the explicit objective of partnering with projects that mainstream gender in their design. Nor did FSI focus on mainstreaming gender in its own operations beyond complying with standard workplace expectations such as non-discrimination and inclusion. Of course, this did not mean the project avoided engagement with gender issues. Nutrition considerations in agriculture research and practice are deeply connected to issues of gender. Food access, food utilisation, control of household resources, division of household labour and women's decision-making autonomy all play a role in mediating food and nutrition security. The FSI partners' strategic interests in improving women's economic empowerment through FSI activities also exposed the team to considerations of gender.

More fundamentally, the methods selected for catalysing innovation through FSI activities had a profound effect on the way the science was conducted. This created a range of opportunities and costs for both women and men who took part in brokering in order to spark food systems innovation. The following section shares some of the challenges and surprises from the perspective of one female scientist attempting to broker for innovation from within a multidisciplinary science organisation.

Brokering for change: implications for gender and science

Brokering as a scientist in a multidisciplinary applied science organisation presents a range of challenges.[2] These challenges are derived from a range of sources – institutional, organisational and individual – and can hamper (or enhance) the brokering experience and its outcomes.

From an institutional perspective, the practice of brokering occurs in a culture where science has dominance over non-science support roles, into which brokering most often falls. Moreover, scientific organisations are often entrenched in institutional culture where disciplines, epistemologies and methodologies compete for primacy. This contest can occur in an environment where traditional, individual-based and highly specialised scientific efforts are celebrated more enthusiastically than experimental, collaborative and applied efforts. Although never explicitly stated, it was not uncommon for this author to encounter views from colleagues which firmly placed process-oriented activities in a category less worthy of scientific pursuit than more traditional expert-driven approaches. Creating legitimacy in a novel area of work will always demand cultural change but, without it, the benefits are less likely to be realised.

A second challenge for this author and other scientists working in FSI was the need to create unconventional products which, among other purposes, served as 'boundary objects' for engaging and facilitating cooperation with development (policy) partners. These products largely took the form of concise syntheses of current research and practice, such as dossiers, practice notes and discussion papers. While they may contain analysis, their purpose was rather to stimulate dialogue on a topic in order to trigger action. Learning how to craft these products took time, time which was not spent crafting manuscripts for publication.

These outputs led to CSIRO scientists being invited by DFAT colleagues to provide scientific input to several strategic development policy and programming initiatives.[3] While these successes might be expected in other areas of science, for an emerging topic like NSA, where experience and capability is still emerging, the opportunity to influence development policy was exciting.

For public research organisations, the evaluation of scientific impact using biblio- and citation metrics is still very much the norm. For organisations which embrace innovative

---

2   The author has written a *Practice Note* on her experiences of brokering as a scientist. This Note can be accessed from <http://partnershipbrokers.org/w/journal/learning-to-work-differently-a-scientists-reflection-on-acting-as-a-broker-for-food-systems-innovation/>.
3   For example, the author contributed to DFAT's new *Strategy for Australia's Investments in Agriculture, Fisheries and Water*, co-authored DFAT's *Operational Guidance Notes on NSA Programming* and was a member of the design team for Timor-Leste's Agricultural Livelihoods Program.

approaches to sustainability challenges, less tangible but no less significant accomplishments need recognition, reward and evaluation. While historically there has been a lack of established organisational incentives to recognise the effort in and impact of successful collaborative processes, changes at CSIRO are afoot.[4] Processes which build cooperation among critical partners, resolve conflict, mediate difficult processes and engage diverse stakeholders and communities are all likely to inspire scientists to diversify their portfolios and skill sets. Adequately resourcing and supporting these new ways of working will no doubt be a future challenge, as will aligning them with formal performance appraisal, promotion and recruitment systems.

Process facilitation techniques such as brokering require possession of special skill sets not necessarily common to scientists. Skills for building trust quickly, communicating cross-culturally and mediating conflict were anticipated. Other skills and traits useful to the brokering process, like inspiring and facilitating participation in innovation forums, organising and managing meetings, and being in possession of a service-delivery mindset, required a shift in attitude, at least for some. The FSI experience enabled a small group of scientists to build on these skills, which ultimately helped to deliver science impact at the policy level.

Gender studies published in business, research and management literature have explored gender differences in several areas relevant to sustainability science including leadership, cooperation and communication (Del Giudice 2015; Jones and Swiss 2014; Castillo 2015; Sanders et al. 2011; van Rijnsoever and Hessels 2011). Findings tend to suggest that both men and women aspire to leadership roles and that both make effective leaders. There are, however, distinct differences in work styles and these have the potential to affect capacity for research collaboration and process facilitation, which are essential to sustainability science approaches. In general, women are more likely to engage in collaborative processes and consult more, and are highly skilled in interpersonal communication and relationship-building. Men tend to promote their points of view in meetings more effectively, are more willing to take risks and are less affected by their emotions at work.

A tempting line of argument might go something like this. Given that women are more likely to possess the soft skills needed for process facilitation, a combination of cultural expectations and social norms help to shape women's participation in activities such as brokering. If organisations fail to formally recognise and reward such activities as critical to achieving impact, the opportunity costs for women researchers to participate in these activities are potentially high. This might lead us to conclude that a gender penalty exists for women scientists practising innovation brokerage in a traditional research setting, since women are disadvantaged by the intense resources brokering consumes and are therefore hampered in their career progression. But is this really the case?

There is no doubt that, for FSI brokers, the time spent on administrative and operational tasks had a considerable impact on scientific output capacity. Tasks such as event management and coordination meant less time was available for practising technical skills and producing traditional science outputs. It is true that more women than men participated in brokering for FSI but it would be premature to conclude that FSI brokers were negatively affected by brokering in relation to formal performance recognition or that their career progression was compromised. Without conducting a qualitative study, draw-

---

4 CSIRO is developing an internal impact evaluation program that aims to build capacity of individuals and the organisation to measure and report on the environmental and societal benefits of innovation. Social impact assessment using theories of change and impact pathways as tools to demonstrate learning and change are examples of approaches under consideration.

ing conclusions on the effects of brokering for women's and men's workplace experience is not possible. Under a different lens, these skill sets might have given women a critical advantage in a professional domain that is increasingly recognised as key to organisational performance.

A different version of the same argument might also hold true. For men, the cost of brokering in research settings is as high or higher since men are less likely to possess the skills required for brokering and are less attracted to acquiring them, given the historic lack of institutional and organisational incentives to pursue process-facilitated activities. Men are, after all, also hampered by organisational barriers such as the lack of incentives for brokering efforts. In addition, gender biases which stereotype women as naturally possessing soft skills may be alienating men from practising innovation brokerage. Given these conditions, the experience of brokering for men in a research organisation may be just as challenging. Of course, gender differences in workplace experiences of brokering might be influenced by other factors such as individual personality traits and project governance.

This is an exciting era of organisational development in CSIRO. A window of opportunity has opened up with the release of *CSIRO Strategy 2020*. The strategy firmly places innovation capacity as integral to achieving science impact and addressing the domestic and international challenges facing Australia. In the midst of brokering, for this author, the costs of brokering and process facilitation have enabled impact of this kind. This author has built trust, fostered relationships and played a role in influencing Australian-sponsored development policy. This author has also uncovered and learned skills along the way. These are just some of the personal and professional benefits presented by brokering for innovation.

Conclusion: brokering for change in a multidisciplinary applied science organisation

Sustainability science, innovation systems thinking and implementation science approaches all have an interest in rendering knowledge usable, applying novel techniques, inspiring experimentation and nurturing the development of a range of capabilities (Kates et al. 2001; Klerkx and Gildemacher 2012). For scientists working on sustainability challenges, especially those who pursue innovation as an impact pathway, brokering to inspire learning and change can be a useful tool. Brokering does, however, raise specific challenges for research organisations and for the women and men who pursue such approaches.

For scientists, brokering requires a shift in mindset on several levels. Expert knowledge becomes one of a range of knowledge types useful for change. Skills in facilitating processes become just as valuable as skills in writing science papers. Learning and action assume importance, and partnerships become catalysts for innovation. For many traditional scientists, these can be considerable personal and professional challenges.

Mutual respect, reciprocity and transparency have guided this innovation broker along the journey. Willingness to compromise, negotiate and remain flexible was critical. Finding comfort in uncertainty and accepting imperfect outputs and processes also helped this broker make ground. Without these ingredients, trust would not have been built as quickly, communication products would have missed their target audience and outcomes would have been harder to achieve.

Brokering is time-intensive. Days and weeks were consumed in conversations, reflection, planning and problem-solving. Creating suitable communication products which helped to facilitate dialogue among diverse stakeholders took considerable time and skill. It is true that these results could instead have been co-produced with science communicators, but their completion by a scientist created legitimacy and built trust between policy and research partners. Their completion also helped this scientist understand and communicate her work more broadly.

There is no doubt of the multiple organisational benefits from integrating gender diversity of thought, experience and perspectives. This is especially true when seeking to address sustainability challenges. Women and men experience work differently and have a range of traits, attitudes and behaviours that potentially facilitate (and hinder) process-oriented activities. For organisations in pursuit of innovation as an impact pathway, investing in capabilities which facilitate integration of knowledge, foster inclusion and stimulate cooperation among diverse stakeholders to negotiate complex processes will be conducive to achieving impact.

Organisational mechanisms which motivate and reward efforts to pursue experimental methods, particularly in the social sciences, are critical for addressing sustainability challenges such as food systems governance. Sustainability science demands new ways of working – without adequate resourcing, rewarding and support, scientists of both genders are disadvantaged.

References

Brooks S, Loevinsohn M (2011) Shaping agricultural innovation systems responsive to food insecurity and climate change. *Natural Resources Forum* **35**, 185–200. doi:10.1111/j.1477-8947.2011.01396.x

Castillo GA (2015) Gender and social research. *International Encyclopaedia of the Social and Behavioural Sciences*. 2nd edn, Vol. 9, pp. 715–722.

Del Giudice M (2015) Gender differences in personality and social behavior. *International Encyclopaedia of the Social and Behavioural Sciences*. 2nd edn, Vol. 9, pp. 750–756.

Foran T, Butler JRA, Williams LJ, Wanjura WJ, Hall A, Carter L, Carberry PS (2014) Taking complexity in food systems seriously: an interdisciplinary analysis. *World Development* **61**, 85–101. doi:10.1016/j.worlddev.2014.03.023

Gibbs EPJ (2014) The evolution of One Health: a decade of progress and challenges for the future. *Veterinary Record* **174**, 85–91. doi:10.1136/vr.g143

Jerneck A, Olsson L, Ness B, Anderberg S, Baier M, Clark E, Hickler T, Hornborg A, Kronsell A, Lövbrand E, Persson J (2011) Structuring sustainability science. *Sustainability Science* **6**, 69–82. doi:10.1007/s11625-010-0117-x

Jones RC, Swiss L (2014) Gendered leadership: the effects of female development agency leaders on foreign aid spending. *Sociological Forum* **29**(3), 571–586. doi:10.1111/socf.12104

Kates RW, Clark WC, Corell R, Hall M, Jaegar CC, Lowe I, McCarthy JJ, Schellnhuber HJ, Bolin B, Dickson NM, Faucheux S, Gallopin GC, Grübler A, Huntley B, Jäger J, Jodha NS, Kasperson RE, Mabogunje A, Matson P, Mooney H, Moore B III, O'Riordan T, Svedin U (2001) Sustainability science. *Science* **292**(5517), 641–642. doi:10.1126/science.1059386

Klerkx L, Gildemacher P (2012) The role of innovation brokers in agricultural innovation systems. *Agricultural Innovation Systems: An Investment Sourcebook*. World Bank, Washington DC.

Lang DJ, Wiek A, Bergmann M, Stauffacher M, Martens P, Moll P, Swilling M, Thomas CJ (2012) Transdisciplinary research in sustainability science: practice, principles, and challenges. *Sustainability Science* **7**(Supplement 1), 25–43. doi:10.1007/s11625-011-0149-x

Njenga M, Karanja N, Prain G, Lee-Smith D, Pidgeon M (2011) Gender mainstreaming in organisational culture and agricultural research processes. *Development in Practice* **21**(3), 379–391. doi:10.1080/09614524.2011.558061

Porter F, Sweetman C (2005) *Mainstreaming Gender in Development: A Critical Review.* (Editorial), Oxfam, 2–10.

Rao A, Kelleher D (2005) Is there life after gender mainstreaming? *Gender and Development* **13**(2), 57–69. doi:10.1080/13552070512331332287

Sanders M, Hrdlicka J, Hellicar M, Cottrell D, Knox J (2011) *What Stops Women from Reaching the Top? Confronting the Tough Issues.* Bain & Co./Chief Executive Women (CEW). <http://www.cew.org.au/wp-content/uploads/2014/12/Action-speaks-louder-than-words-CEO-conduct-that-counts.pdf>.

UN Entity for Gender Equality and the Empowerment of Women (2014) *World Survey on the Role of Women in Development: Gender Equality and Sustainable Development.* UN Women, New York.

van Rijnsoever FJ, Hessels LK (2011) Factors associated with disciplinary and interdisciplinary research collaboration. *Research Policy* **40**, 463–472. doi:10.1016/j.respol.2010.11.001

Wittmayer JM, Schäpke N (2014) Action, research and participation: roles of researchers in sustainability transitions. *Sustainability Science* **9**, 483–496. doi:10.1007/s11625-014-0258-4

12

Risk, sustainability and time: sociological perspectives

Stewart Lockie and Catherine Mei Ling Wong

Sustainability is an inherently temporal concept. It is as much about the tempo of social and environmental change and the rights of future generations as it is about the magnitude of change and its implications for human well-being. We do not expect our communities, environments or economic systems to remain unchanged over time, but we do want to maintain or improve their ability to keep on delivering the things we value well into the future.

Of course, moving from general agreement about the importance of long-term sustainability to concrete decisions in the here-and-now about resource use, conservation, pollution standards, spatial planning, infrastructure investment etc. presents several challenges. Not least among these are the challenges of accounting for uncertainty and ignorance. Sustainability is also about risk. It is about accepting that things may not turn out as planned. It is about understanding and managing the potential for things to go wrong, and it is about preparing, despite our best efforts, for things to go wrong in ways that are currently unknowable or unforeseen.

Like so many aspects of sustainability, managing risk requires a genuinely multidisciplinary approach. As a concept that embodies, in simplest terms, the probability and consequences of adverse events, there is a clear role for the actuarial and technical disciplines in understanding and monitoring the causal pathways that lead to and from harmful events – whether these be industrial accidents, natural disasters, long-term decline in the health or availability of natural resources, or any other threat to human well-being. Critically, understanding and managing such risks also raises questions for the social sciences – questions about acceptability, distribution, perception, communication, organisational design and so on.

The question of acceptability alone is enough to demonstrate that risk is a deeply social and political issue. No matter how accurately we can define the probability of an adverse event, only people can decide whether the risk, as understood, is one worth taking. Only people can decide whether that risk is more or less acceptable than other risks, perhaps of a different nature or characterised by more or less uncertainty. Take, for example, the potential for catastrophic failure in hazardous facilities such as nuclear power plants and the very different kinds of risk associated with anthropogenic climate change. Should Japan be restarting its nuclear industry with revised safeguards in the wake of the 2011 Fukushima Daiichi disaster? Is this more or less acceptable a risk than those posed by

Japan's substitution of nuclear power generation with comparatively carbon-intensive fossil fuel-based generation?

The more we understand the likelihood and potential consequences of particular risk events, the more productive political debate over their relative acceptability is likely to be. However, the social dimensions of risk go beyond our collective response to technically defined estimates of probability and consequence. The likelihood and consequences of adverse events are themselves shaped by the ways in which people and institutions perceive and behave in relation to risk. The burden of adverse events tends to fall more heavily on some people than others. Neither the benefits of risk-taking nor exposure to the consequences are equally distributed.

In this chapter, we review what has been learned about the social and political dimensions of risk and what these mean for sustainability researchers and practitioners. In particular, we identify where an overtly sociological approach to risk calculation is warranted, and where emphasis needs be laid in relation to more widespread participation in risk governance.

The limits of technical risk calculation

The industrial age has fundamentally transformed human societies and the ecosystem processes on which we rely. Threats characteristic of pre-industrial times have typically been mitigated, and new threats have emerged. At issue here is not whether contemporary risks are more or less acute than those faced by our forebears. The issue, rather, is how societies are organised to both produce and manage risk. Social theorist Ulrich Beck (1992) famously argued that conflict over the distribution of exposure to risk has, in fact, joined conflict over the distribution of wealth as a defining feature of contemporary society.

Some risks are, of course, well understood and generally well managed with mature processes and institutional arrangements in place for their assessment and mitigation. Insurance, financial instruments, workplace health and safety procedures, emergency services etc. do an excellent job of protecting most of us, most of the time, from the hazards of daily life. However, as anthropogenic climate change attests, human intervention in the global environment has produced unintended consequences of potentially unlimited duration and reach (Giddens 1994). The potential for interactive effects and feedback loops to increase the rate of change or to produce entirely new hazards adds more layers of complexity. Global environmental change is thus exceptionally well understood in some senses and very poorly understood in others – a feature routinely seized upon by opponents of concerted mitigation efforts. Global change confronts established sources of political and scientific authority and the techniques they use to assign responsibility and liability (Beck 1992).

It is hard to dispute the broad generalisation that technical risk calculation and management is challenged by complexity and uncertainty. By itself, however, this provides little guidance to governments and others in relation to how they might actually deal with complexity and uncertainty. As a first step, it is useful to unpack in a little more detail the specific kinds of situations in which calculating and responding to risk in a straightforwardly technical manner may not be possible. According to Measham and Lockie (2012b, pp. 3–5), these include situations characterised by:

- high levels of uncertainty about risk due to, for example, inadequate baseline knowledge, scientific conflict and/or poorly understood causal relationships;
- residual uncertainty about otherwise well understood risks at other spatial or temporal scales;

- value conflicts in relation to the acceptability of risk exposures and mitigation strategies – such conflicts being particularly evident in situations characterised by:
  - hazards that are extremely low in probability but which have potentially catastrophic consequences;
  - an unequal distribution among stakeholders of the costs and benefits arising either from risk-inducing activities or from risk reduction and management activities;
  - a lack of confidence in the capacity or trustworthiness of expert and/or risk regulating institutions;
- interactive and dynamic relationships between risk assessment and risk management.

It could be argued that the last of these points applies to all risk scenarios; that is, that the objective level of risk posed by any identifiable threat is related directly to the effectiveness of measures taken to mitigate and/or avoid that threat. It follows that risk assessment necessarily makes assumptions about how people and institutions will behave in response to risk, the effectiveness of risk mitigation measures and so on. It does not follow, however, that these assumptions necessarily introduce sufficient uncertainty to undermine the validity of technical risk assessments. The question is whether the assumptions made about people, institutions and, for that matter, technology are robust and, if not, what can be done to improve them?

Failure to reflect critically on the social assumptions built into calculations of risk can lead risk assessors and managers to ignore the impact of risk perception, management and other human factors on risk-taking behaviours and accident rates, and to ignore opportunities to further reduce those risks they have deemed acceptable (Hopkins 2005). Before discussing how and when a more sociological approach to risk calculation might be taken, we will examine in more detail the issues of distribution, institutional capacity and trust identified by Measham and Lockie (2012a) as distinctly social elements of risk.

Distribution of risks and benefits

Controversies around risk often reveal inequalities in exposure, vulnerability and resilience to risks. The most vulnerable tend to also be those most exposed to risks. The poor are often less resilient to risk. They live in places where infrastructure is weak and access to basic health services and utilities are inadequate, increasing their risk of death, loss of livelihood and so on. Racial minorities and the poor, for example, were far more affected by Hurricane Katrina and took much longer to recover from the disaster. No doubt the wealthy neighbourhoods of New Orleans also suffered impacts but, being located at higher elevations, 'White Teapot' districts like Uptown, Carrollton, University and the French Quarter were far less exposed to storm surges from the hurricane and to the chemically contaminated floodwaters that followed. By contrast, the neighbourhoods at lower elevations were largely occupied by African Americans and had poor access to transportation. This increased the residents' vulnerability to hurricane surges, hampered their ability to seek refuge and posed greater challenges to their capacity to recover. Over a week after the hurricane, a significantly greater percentage of African American residences remained flooded in the metropolitan New Orleans area compared to those of other ethnic groups (Morse 2008). Chemical contamination was a grave concern in these areas, but it was difficult to assess the impact of the hurricane because pre-disaster contamination levels were already above government standards. About 40% of soils in the city exceeded safe levels of contamination, and 20–30% of inner-city children were already living with blood lead levels in excess of the health guidelines set by the Center for Disease Control and Prevention (Morse 2008).

Polluting industries also tend to be located in poor areas where regulation and/or enforcement are comparatively weak. This often results in either the involuntary residence of ethnic minorities or marginal communities in polluted industrial towns due to the lack of mobility, or the involuntary relocation (i.e. displacement) of communities from their (traditional) lands to make way for industrial development. The agricultural office of Union Carbide India Ltd (UCIL), responsible for the 1984 Bhopal disaster, for example, was moved from Bombay to Bhopal in 1968. This was meant to generate development in Madhya Pradesh, a poor province by Indian standards. It also opened UCIL operations to local corruption and a structural inability to handle hazardous technology. When the gas leak broke, the first people exposed were those living in the slums next to the pesticide plant (Varma and Varma 2005). Within the first week of the disaster, half the population fled the city – those who remained were too weak or too poor.

Risk and organisations

Catastrophic industrial accidents are most commonly described as technological disasters. However, there is ample evidence to suggest that, in most cases, the primary causes of catastrophic industrial accidents are neither technological failures nor the actions of people operating the relevant facilities at the time of the accident. The primary causes can almost always be traced back to failures in the organisations that design, operate and regulate hazardous facilities.

In making this claim, we do not dispute that some risks cannot be adequately mitigated or managed. No matter how well designed or diligently regulated, there are any number of risks simply not worth taking. This leads some social scientists to argue that catastrophic accidents are inevitable in all hazardous industries because of the complexity of technological systems and the difficult environments in which they often operate. Despite their intuitive appeal, however, such arguments fail to explain accidents in sociological terms or to offer additional insights into what official inquiries can already tell us about the causes of specific events (Hopkins 2005). In fact, official inquiries consistently identify a small list of social and institutional factors responsible for creating the conditions under which often small, possibly even routine, component failures within technological systems have initiated chains of events that eventually result in catastrophic system failure. These factors include:

- poor regulation;
- ignored warning signs;
- production pressures;
- cost-cutting;
- inadequate training.

We will examine regulation in the following section. Meanwhile, it is important not to interpret the rest of the above factors as simply examples of operator error. As Perrow (2011) pointed out, it is a rare person who never makes mistakes or bends rules, but the consequences of these mistakes more often become catastrophic when they are magnified by organisations. It is organisations that assemble the technological infrastructure and materials that are capable of causing catastrophic industrial accidents (Perrow 2011), and it is organisations that institutionalise incentive systems which prioritise production and profitability over safety (Hopkins 2005). It is organisations that marginalise dedicated safety personnel and processes through their authority and reporting structures, and it is organisations that underinvest in maintenance and training.

Vaughan's (1996) analysis of the Space Shuttle Challenger disaster, for example, contradicted conventional explanations that attributed the disaster to managerial wrongdoing. Certainly, had the decision been left solely to engineers responsible for risk assessment the launch would have been cancelled. Instead, managers pushed ahead against engineering advice. Nonetheless, Vaughan argued that mistakes, mishaps and the potential for disaster had been building within the organisational structures of NASA since the 1970s. As production pressures intensified within NASA, organisational goals and internal norms started to change. Warnings written into engineering risk assessments were repeatedly downplayed, leading over time to a shift in the way technical information was interpreted by decision-makers and the transformation of unsafe practices into acceptable behaviour. Financial scarcity and competition for resources drove the suppression of information about safety violations and undermined scientific debate. This impeded NASA's ability to learn from past mistakes and helped instead to normalise error.

Similarly, the Macondo oil rig in the Gulf of Mexico was six weeks behind schedule and under pressure from management at BP headquarters to speed up drilling (Barstow et al. 2010). This placed additional pressure on defence in depth systems built into the rig, and warning signals that were issued by the safety systems were ignored by operators. Confirmation bias (Nickerson 1998) told operators that warning signals were normal, a reflection of faulty instruments or testing procedures, and nothing to be unduly concerned about. Post-disaster investigations concluded that the conduct of safety tests had been little more than a 'box-ticking' exercise. Operators carrying out integrity tests of wells had done so with the assumption that earlier stages of the job had been done properly, leading them to reconceptualise indicators of partial malfunction as normal (Hopkins 2012). In sum, technological systems routinely fail when the prioritisation of production over safety sees operations accelerated, maintenance delayed, backup safety systems decommissioned, and disregard for warning signs institutionalised.

Mistakes can be institutionalised even before potentially hazardous activities become operational. The Fukushima Daiichi plant, for example, was designed against what was believed to be a relatively conservative set of safety parameters – built to withstand a magnitude 8.6 earthquake and waves up to 5.7 m. This proved inadequate to deal with the magnitude 9 earthquake and 14 m waves that hit the coast. Subsequently, it was found that historical records of tsunamis in Japan were measured with only moderate accuracy and were not well documented (Hagmann 2012). As the previous section argued, sole reliance on technical risk assessment can lead to systematic underestimation of the likelihood of an accident and failure to identify opportunities to reduce that risk. Planning processes must account for potential underestimation, given that calculation of probabilities necessarily depends on incomplete information (Taleb 2007).

It is important to note that the institutionalisation of safety procedures is not, by itself, sufficient to avert catastrophic accidents (Power 2004; Reason 2000). A survey of the US nuclear industry in the 1980s, for example, found that 60% of all human performance problems could be attributed to procedures that were unworkable, incomprehensible or simply wrong (INPO 1985 from Reason 2000). Safety procedures without context can sometimes be disastrous. In the 1988 North Sea gas and oil platform explosion, most of the 165 rig workers who died complied strictly with the safety drill by assembling at the accommodation area – which was directly in line with a subsequent explosion (Reason 2000). Decisions made by office-based middle- to upper-level management can be out of sync with frontline operational realities (Hopkins 2005).

The idea of safety culture has been proposed as a way to address institutional weaknesses and the apparent inability of some organisations to learn and respond flexibly to emergent threats. According to Reason (1997), safety culture comprises four key components including a:

- reporting culture, in which people are prepared and encouraged to report their errors and near-misses;
- just culture, where people are encouraged and even rewarded for providing essential safety-related information;
- flexible culture, in which conventional hierarchical modes of decision-making can be shifted to a flatter structure where control passes to task experts during an emergency and reverts to the traditional bureaucratic mode once the emergency has passed;
- learning culture, in order to inculcate the willingness and competence to implement major reforms when needed.

High Reliability Organization (HRO) research similarly identifies several features of organisations that operate in potentially hazardous sectors but have managed to avoid major accidents. These features include clear and consistent goals that ensure reliable operations, organisational leadership that prioritises safety in operations, redundancy in technical operations (i.e. backup safety systems), continuous training, direct lines of communication, hierarchical differentiation and a culture that instils accountability and responsibility (Ramana 2012; Roberts 1990). It is the collective mindfulness of the organisation as a whole which facilitates the discovery of the unexpected and the correction of errors before they escalate into catastrophe (Weick *et al.* 1999).

Both safety culture and HRO attempt to internalise a system of reflexivity within organisations – one using cultural sensibilities and the other by organisational design. However, it is important not to treat ideas like 'safety culture' as silver bullet solutions to risk. As Pidgeon and O'Leary (2000) note, organisations are notoriously unreflexive and resistant to learning and reform. Internally, they tend to lack the required (institutional and financial) motivation and levels of organisational self-awareness to conduct serious self-examination (Flynn 2003). This is, in part, because of the ever-present incentive to prioritise profitability and efficiency over the more nebulous reward of accidents averted. Some organisations in high-risk industries such as nuclear power, oil and gas attempt to change the incentive structure to augment good safety performance by creating financial and symbolic rewards, yet these measures remain relatively weak and can lead to under-reporting of minor incidents and near-misses that can build up to major events.

Regulatory institutions

Regulatory systems are designed to avert some of the organisational weaknesses discussed here. They embody the rules and norms that keep industries in check through the administration of legal requirements such as the requirement that industries conduct environmental and social impact assessments, comply with licensing requirements and so on. In principle, regulators serve the interest of public health and safety. They typically report to the government and function independently of industry.

Regulatory arrangements, however, sometimes fall prey to regulatory capture. This is when the regulator loses autonomy to independently evaluate the industry and instead modifies regulations and their implementation to suit industry interests. This can occur from the outset, when financial and/or administrative arrangements place the regulator in a dependent relationship with the industry it is meant to regulate. It can also develop over time as a consequence of 'revolving door' employment patterns; that is, the frequent circulation of personnel between industry and regulatory bodies. Such exchanges can be used by regulated firms to assert pressure within the regulator by offering agency staff lucrative employment opportunities in exchange for being cooperative and by inducing the regulators to identify with the regulated industry (Hardy 2006).

One case in point is the Atomic Energy Regulatory Board (AERB) in India. Officially, the regulator has responsibility for the industry, overseeing the Department of Atomic Energy (DAE) which is accountable for the nuclear industry. But the AERB reports to the Atomic Energy Commission, whose Chairman is also the Secretary of the DAE. This allows the DAE to exercise administrative powers over the regulator. Furthermore, 95% of the AERB's evaluation committees are scientists and engineers on the payrolls of the DAE. Similar instances of regulatory capture were found in the US nuclear industry, to which the Three Mile Island nuclear accident in 1979 was partly attributed (Rees 1994). Some three decades later, the same conclusions were drawn by the National Diet of Japan in its independent investigation of the Fukushima Daiichi nuclear disaster. In what it called a man-made disaster, the Diet found that collusion between the government, regulators and TEPCO (operator of the nuclear plant) supported faulty rationales for decisions and actions which led to the 2011 catastrophe (National Diet of Japan 2012).

Addressing regulatory capture, like other aspects of risk aversion, requires an understanding of the social and institutional context for regulation; in this case, the governance of risk through a variety of interacting institutions. More participatory approaches to risk assessment and regulation are thereby advocated by several authors in order to improve transparency and accountability, utilise citizen knowledge and expertise, and facilitate communication and learning among multiple agencies (Lockie 2001; Renn 2015; Wong 2015). These are discussed in more detail below.

Trust and social amplification of risk

When organisations fail their duty of care, public distrust can be both a source and outcome of risk. Studies have found a strong correlation between public trust in the organisations tasked to manage risk and public risk perceptions. They posit that when trust is strong, risk perceptions tend to be low. Conversely, distrust in organisations can lead people to oppose risk even when it is perceived to be low (Freudenburg 2003; Siegrist and Cvetkovich 2000). The failure of organisations to carry out their responsibilities to the broader public therefore has amplifying effects on actual risk. The 'principle of asymmetry' suggests that trust is much more difficult to build or rebuild because trust-destroying negative events carry much greater weight than positive events (Slovic 1993). The lack of transparency, false reporting, regulatory capture and institutional body language of scientific organisations that denigrate lay knowledge or public concerns (Wynne 1992) all contribute to public distrust in organisations, with amplifying consequences.

Risk communication has therefore become a central focus in risk research. In trying to explain why some risk events have far-reaching effects on communities, industries and countries that are seemingly unrelated, the Social Risk Amplification Framework (SARF) focuses on how information flows and transforms as it passes through communication stations. SARF argues that social agents generate, receive, interpret and pass on risk signals. In the process, signals get transformed as they filter through various social and individual amplification stations which can increase (amplify) or decrease (attenuate) the volume of information about an event, heighten the salience of certain aspects of a message or reinterpret and elaborate the available symbols and images. This sends further signals to other participants in the social system, which can produce secondary and tertiary ripple effects that spread beyond the initial impact of an event into other previously unrelated technologies, industries and institutions (Kasperson et al. 2010), often with lasting effects. The Bhopal disaster, for example, happened in the chemical industry in 1984 and had a huge impact on public trust in the regulatory system and large corporations. More than two decades later, associations were made between the Bhopal tragedy and nuclear development in India, which galvanised public resistance and became a key driver for legislative reform in the *Civil Nuclear Liability Act 2010*. The new law marked a departure from international norms: it placed financial liability for an accident in India on the (foreign) supplier company instead of the (domestic) operator, which in turn stalled major civilian nuclear negotiations with the US, France and Japan.

Social amplification of risks is more than a matter of perception. It produces very material consequences for individuals and their families. The Deepwater Horizon accident, for example, was responsible not only for the deaths of 11 people on the rig but for a range of impacts on coastal residents from Texas and Louisiana, to Mississippi, Alabama and Florida as a consequence of the oil spill that followed. Many of the coastal residents depended on local seafood harvesting for commerce, recreation and subsistence. It was the basis of their regional economy. But perceived contamination of local seafood led to a slump in sales, which in turn had social and psychological impacts on local residents. Higher than expected levels of mental illness, substance abuse and family strife were recorded in the aftermath of the spill (Kane 2015).

Socially informed risk assessment and management

Risk may well be inherently social, but it would be neither practical nor helpful to suppose that every exercise in risk assessment must be multidisciplinary and/or participatory. The question then becomes how sustainability professionals and others might make informed decisions about when to utilise technical risk assessment alone and when to take a more expansive approach.

Renn and Klinke (2012) utilised the concept of the risk escalator to illustrate when technical risk assessment is sufficient, and thence the circumstances under which additional approaches and stakeholders ought to be engaged (see Fig. 12.1). The risk escalator model is based on clear distinctions between the concepts of complexity, uncertainty and ambiguity:

- complexity – difficulty identifying and quantifying causal links between a multitude of potential candidates and specific adverse effects;
- uncertainty – the limitation or absence of scientific knowledge (data, information) that makes it difficult to assess the probability and possible outcomes of undesired effects;
- ambiguity – a situation of ambivalence in which different and sometimes divergent streams of thinking and interpretation about the same risk phenomena and their circumstances are apparent.

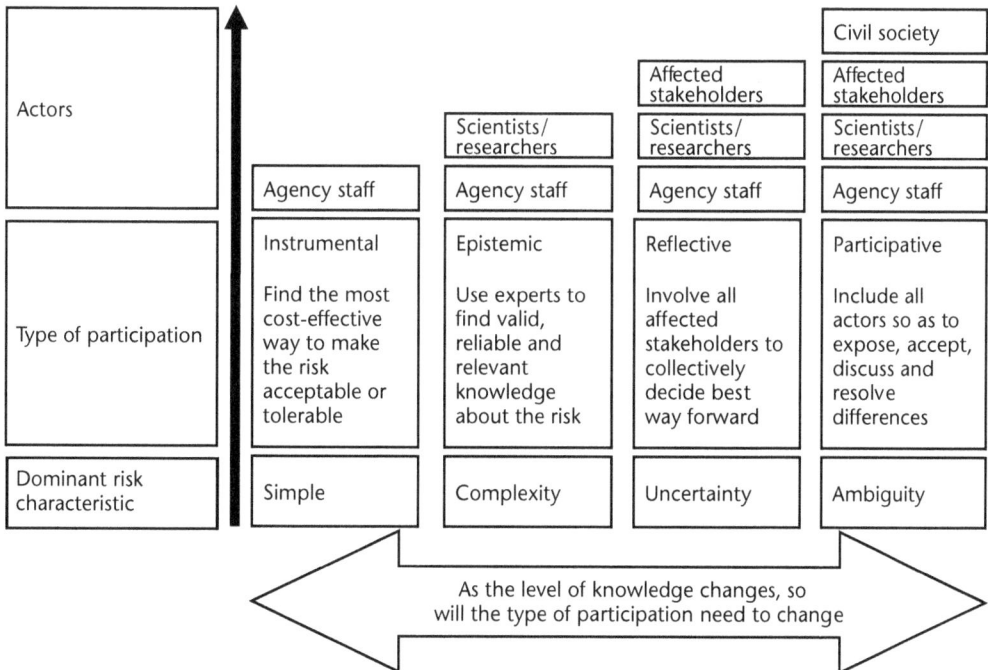

Fig. 12.1. The risk escalator. Different stakeholder involvement requirements for risks characterised by simplicity, complexity, uncertainty and politico-moral ambiguity. Source: Renn and Klinke (2012, p. 54).

In turn, there are two kinds of ambiguity:

- interpretative ambiguity – the variability of interpretations based on identical observations or data assessment results;
- normative ambiguity – different concepts of what can be regarded as tolerable, e.g. ethics, quality of life parameters, and distribution of risks and benefits.

Risk problems most in need of a more expansive approach to decision-making often embody a combination of complexity, uncertainty and ambiguity. They also tend to evolve over time, as does public acceptability (or otherwise). Nuclear power in Japan, for example, was largely considered an acceptable risk by the public until the Fukushima Daiichi disaster. After the tragedy, the risk issue became bigger than a simple matter of technical safety. It was more about organisational failure, the social structure of the industry, and political collusion between the nuclear power corporation and the Japanese government.

The challenge, therefore, is how to go about implementing socially informed risk assessment that is amenable to the evolution of risk problems and its social acceptability. Deliberative approaches do well to facilitate collective choice through mechanisms such as deliberative polls, citizens' juries, planning cells and consensus forums. These platforms provide a way for decision-makers to engage with public stakeholders, identify collective interests and shared values, account for societal values, and make decisions in a collaborative and transparent manner (Dryzek 2004). They, however, often lack a set of practicable tools and processes for more substantive participation by stakeholders throughout the lifecycle of a project.

In an attempt to address this limitation and the unstable nature of risk problems, a hybrid risk governance framework has been proposed as a way to organise multi-stakeholder participation from the planning and design stages of a project through to implementation and regulation (Wong 2015). This format of collective decision-making is done in five stages, with a reflexive component at each stage to allow for adjustments and corrections as the project develops.

- Stage 1: Co-selection – collective selection of participants in the deliberative process.
- Stage 2: Co-design – identify and align goals and interests.
- Stage 3: Co-definition – integrate lay knowledge, social values and public preferences in risk estimation, evaluation and management plans.
- Stage 4: Co-planning – develop shared responsibilities and resources.
- Stage 5: Co-regulation – allocate roles and identify deliverables.

Conclusion

There will always be conflict over the acceptability and management of risk, especially when complex and ambiguous risks evoke competing interpretations, interests and values. As we have seen, however, there is considerable potential for more sociologically sophisticated approaches to risk calculation and governance to:

- reduce the objective likelihood of adverse events through improved understanding of how people behave in relation to risk and systematic analysis of the institutions that design, operate and regulate hazardous activities;
- improve the effectiveness of risk regulation and management by creating avenues for multidisciplinary and public oversight of the relationships between regulators and those who undertake potentially hazardous activities;
- promote more equitable distributions of risk exposure and benefits by making existing and potential injustices more visible and thus viable targets of policy, dialogue and decision-making.

Participatory approaches, similarly, have considerable potential to:

- improve decision-making and reduce uncertainty by increasing the pool of knowledge available to all stakeholders in potential risk events;
- better account for social values/preferences in decisions, especially when risk problems are complex, uncertain and/or ambiguous;
- enhance transparency and build trust;
- promote social learning and thus assist decision-makers to address risk as an ongoing and evolving process.

People often forget to be afraid, but it isn't required of individuals to live in constant states of fear or hypervigilance in order to avert adverse events. Sitting behind the technological systems, ecosystem interventions, spatial development patterns and so on that produce distinctively modern risks are a host of institutional arrangements intended both to manage risk on our behalf and to provide opportunities for deliberation over what risks ought to be deemed acceptable. Such arrangements are necessarily imperfect. But before falling back on the circular reasoning behind claims that catastrophic outcomes are simply an inevitable consequence of the industrial age, we should look to what all the sciences can tell us about the production and mitigation of risk. Sociological perspectives bring the imperfection of institutional arrangements to the fore; in so doing, they open new ways of thinking and doing something about risk and sustainability.

References

Barstow D, Rohde D, Saul S (2010) Deepwater Horizon's final hours. *New York Times*, 25 December. <http://www.nytimes.com/2010/12/26/us/26spill.html?pagewanted=all&_r=0>.

Beck U (1992) *Risk Society: Towards a New Modernity*. Sage, London.

Dryzek JS (2004) Pragmatism and democracy: in search of deliberative publics. *Journal of Speculative Philosophy* **18**(1), 72–79. doi:10.1353/jsp.2004.0003

Flynn J (2003) Nuclear stigma. In *The Social Amplification of Risk*. (Eds N Pidgeon, RE Kasperson, P Slovic) pp. 326–352. Cambridge University Press, Cambridge.

Freudenburg W (2003) Institutional failure and the organizational amplification of risks: the need for a closer look. In *The Social Amplification of Risk*. (Eds N Pidgeon, RE Kasperson, P Slovic) pp. 102–120. Cambridge University Press, Cambridge.

Giddens A (1994) Living in a post-traditional society. In *Reflexive Modernization: Politics, Tradition and Aesthetics in the Modern Social Order*. (Eds U Beck, A Giddens, S Lash) pp. 56–109. Stanford University Press, Stanford, CA.

Hagmann J (2012) Fukushima: probing the analytical and epistemological limits of risk analysis. *Journal of Risk Research* **15**(7), 801–815. doi:10.1080/13669877.2012.657223

Hardy DCL (2006) *Regulatory Capture in Banking*. Working Paper WP/06/34. International Monetary Fund. <https://www.imf.org/external/pubs/ft/wp/2006/wp0634.pdf>.

Hopkins A (2005) *Safety, Culture and Risk: The Organisational Causes of Disasters*. CCH Australia, Sydney.

Hopkins A (2012) *Disastrous Decisions: The Human and Organisational Causes of the Gulf of Mexico Blowout*. CCH Australia, Sydney.

Kane AS (2015) Five years after the Deepwater Horizon oil spill: impacts on Gulf communities and seafood. *The Conversation*, 20 April. <http://theconversation.com/five-years-after-the-deepwater-horizon-oil-spill-impacts-on-gulf-communities-and-seafood-40138>.

Kasperson JX, Kasperson RE, Pidgeon NF, Slovic P (2010) The social amplification of risk: assessing 15 years of research and theory. In *The Feeling of Risk: New Perspectives on Risk Perception*. (Ed. P Slovic) pp. 317–344. Routledge, London.

Lockie S (2001) SIA in review: setting the agenda for impact assessment in the 21st century. *Impact Assessment and Project Appraisal* **19**(4), 277–287. doi:10.3152/147154601781766952

Measham T, Lockie S (2012a) *Risk and Social Theory in Environmental Management*. CSIRO Publishing, Melbourne.

Measham T, Lockie S (2012b) Social perspectives on risk and uncertainty: reconciling the spectacular and the mundane. In *Risk and Social Theory in Environmental Management*. (Eds T Measham, S Lockie) pp. 1–14. CSIRO Publishing, Melbourne.

Morse R (2008) *Environmental Justice Through the Eye of Hurricane Katrina*. Joint Center for Political and Economic Studies, Health Policy Institute, Washington DC.

Nickerson RS (1998) Confirmation bias: a ubiquitous phenomenon in many guises. *Review of General Psychology* **2**(2), 175–220. doi:10.1037/1089-2680.2.2.175

Perrow C (2011) Fukushima and the inevitability of accidents. *Bulletin of the Atomic Scientists* **67**(6), 44–52. doi:10.1177/0096340211426395

Pidgeon N, O'Leary M (2000) Man-made disasters: why technology and organizations (sometimes) fail. *Safety Science* **34**(1–3), 15–30. doi:10.1016/S0925-7535(00)00004-7

Power M (2004) The risk management of everything. *Journal of Risk Finance* **5**(3), 58–65. doi:10.1108/eb023001

Ramana MV (2012) *The Power of Promise: Examining Nuclear Energy in India*. Penguin, London.

Reason J (1997) *Managing the Risks of Organizational Accidents*. Ashgate Publishing, Sydney.

Reason J (2000) Safety paradoxes and safety culture. *Injury Control and Safety Promotion* **7**(1), 3–14. doi:10.1076/1566-0974(200003)7:1;1-V;FT003

Rees JV (1994) *Hostages of Each Other: The Transformation of Nuclear Safety since Three Mile Island*. University of Chicago Press, Chicago.

Renn O (2015) Stakeholder and public involvement in risk governance. *International Journal of Disaster Risk Science* **6**(1), 8–20. doi:10.1007/s13753-015-0037-6

Renn O, Klinke A (2012) Complexity, uncertainty and ambiguity in inclusive risk governance. In *Risk and Social Theory in Environmental Management*. (Eds T Measham, S Lockie) pp. 59–76. CSIRO Publishing, Melbourne.

Roberts KH (1990) Some characteristics of one type of high reliability organization. *Organization Science* **1**(2), 160–176. doi:10.1287/orsc.1.2.160

Siegrist M, Cvetkovich G (2000) Perception of hazards: the role of social trust and knowledge. *Risk Analysis* **20**(5), 713–720. doi:10.1111/0272-4332.205064

Slovic P (1993) Perceived risk, trust, and democracy. *Risk Analysis* **13**(6), 675–682. doi:10.1111/j.1539-6924.1993.tb01329.x

Taleb NN (2007) *The Black Swan: The Impact of the Highly Improbable*. Random House, New York.

National Diet of Japan (2012) *Official Report of the Fukushima Nuclear Accident Independent Investigation Commission*. <http://cryptome.org/2012/07/daiichi-naiic.pdf>.

Varma R, Varma DR (2005) The Bhopal disaster of 1984. *Bulletin of Science, Technology and Society* **25**(1), 37–45. doi:10.1177/0270467604273822

Vaughan D (1996) *The Challenger Launch Decision: Risky Technology, Culture, and Deviance at NASA*. University of Chicago Press, Chicago.

Weick KE, Sutcliffe KM, Obstfeld D (1999) Organizing for high reliability: processes of collective mindfulness. In *Research in Organizational Behavior*. (Eds RS Sutton, BM Staw) Vol. 1, pp. 81–123. JAI Press, Stamford, CT.

Wong CML (2015) The mutable nature of risk and acceptability: a hybrid risk governance framework. *Risk Analysis* **35**, 1969–1982. doi:10.1111/risa.12429

Wynne B (1992) Misunderstood misunderstanding: social identities and public uptake of science. *Public Understanding of Science (Bristol, England)* **1**, 281–304. doi:10.1088/0963-6625/1/3/004

13

Policy-relevant research: improving the value and impact of the social sciences

Brian W. Head

This chapter considers the argument that social science research on sustainability can and should be more influential in public policy. The social sciences are essential because all analyses of socio-ecological problems and preferred ways to address such problems are based on assumptions about human behaviour and choices. The need for individual and collective action is interpreted through the lens of diverse socio-economic needs and interests, which influence the ways in which governance processes shape policies and programs.

It is widely agreed that research about sustainability is driven by concerns to solve real-world problems of sustainability (Kates *et al.* 2001), rather than simply to pursue more traditional models for understanding key features of the physical and bio-social worlds. Sustainability science is inherently linked to promoting sustainability goals, rather than claiming to be detached or neutral (Sayer 2011). Thus, in considering patterns, trends and causal relationships in socio-ecological systems, the ultimate concern is with developing more effective strategies and interventions.

Translating research findings into effective action underlies the motivation of sustainability research communities around the globe. As Kates has proposed, the sustainability sciences address the 'interactions' between social and natural systems and their 'impact' on the challenge of sustainability, namely, the challenge of:

> *meeting the needs of present and future generations while substantially reducing poverty and conserving the planet's life support systems ... sustainability science is a different kind of science that is primarily use-inspired, as are agricultural and health sciences, with significant fundamental and applied knowledge components, and commitment to moving such knowledge into societal action (Kates 2011, pp. 19449–19450).*

To tackle the grand challenges of sustainability, a broad coalition of expertise is needed. Palmer urges the importance of researchers seeking to interact with decision-makers in order to identify shared perspectives on key science problems, and the need for social and biophysical scientists to collaborate in developing and addressing these issues. This type of science can be called 'actionable' because:

it has the potential to inform decisions (in government, business, and the household), to improve the design or implementation of public policies, or to influence public- or private-sector strategies, planning and behaviors that affect the environment (Palmer 2012, p. 6).

The relevance of the social sciences to issues of socio-ecological sustainability has been clear for many decades. For example, in 1989 in the US the National Research Council established a committee on the human dimensions of global change. Its report mapped out a major research program and called for joint action and mutual understanding across the social and environmental sciences. The social sciences have special relevance for understanding the drivers of global change that impact on ecological health (e.g. population, economic growth, technological change), the interaction between local, regional and national scales, and the range of institutional and attitudinal responses to environmental risks (Stern *et al.* 1992).

This chapter focuses on aspirations to increase the relevance and impact of social science contributions to the policy process. Some of the messages are directed towards social science researchers, especially those who puzzle over how to be more influential in a world where the biophysical sciences have precedence. Other messages are directed towards the biophysical scientists who are frustrated as to why their solutions are not being implemented with enthusiasm by decision-makers.

For many decades, one of the underlying goals of policy-engaged scientific research has been to establish processes for scientifically informed policy-making, otherwise described as evidence-based policy. How realistic is this aspiration? And more broadly, in demanding a policy-making system that takes account of best-available evidence, what are the forms of public value offered by the social sciences? To assess these issues, this chapter is divided into three sections:

- first, it provides a brief account of the policy process, focusing on the conflicting types of expertise and practical experience which contend for attention in policy debate and the policy development process;
- second, it explores the various roles available for social science experts wishing to influence the policy process (e.g. as expert advisers within the policy system, or as external lobbyists and publicists on specific issues);
- third, it argues that we should not underestimate the influence of the social science contribution to informed discussion, through various mechanisms of education and training, participation in knowledge-brokering partnerships and raising the quality of public debate.

In a short concluding section, the chapter briefly highlights the underlying basic requirements for providing trusted advice and gaining policy influence. It argues that one of the invaluable roles of social science is in building stakeholder understanding and capability, and in this way contributing to improved implementation of well-crafted initiatives.

The policy process as a contest of ideas

In democratic societies such as the OECD group of countries, there is widespread rhetoric about the importance of evidence-based approaches to policy-making. However, political factors often seem to dominate in framing the problems and shaping the solutions (Weiss and Bucuvalas 1980; Oreskes 2004; Head 2015).

This is a source of great frustration for those who subscribe to the positivist doctrine of the hierarchy of scientific knowledge. That latter doctrine has two components: the claim that evidence drawn from rigorous experimental studies is superior to other forms of knowledge, and the claim that rigorous scientific findings should drive public decision-making. Much of the recent literature on the relationships between science and policy calls into question both of these propositions. Leaving aside the complex philosophical arguments around epistemology and theories of knowledge, the practical limitations of the positivist science-driven approach are evident.

First, in practice, there will always be gaps and uncertainties in knowledge, whether about social, natural or socio-ecological processes. Our knowledge is fragmented, pluralist and provisional (Lindblom and Cohen 1979). In other words, there are few emerging issues of great importance for human well-being and ecological health where the science is comprehensive and compelling. This is especially true of 'wicked' problems, where values and perspectives seem to be in a permanent state of conflict (Head and Alford 2015; Head 2014a, b). In the meantime, in the absence of full information, governments and commercial investors are required to make decisions. In some cases, these involve fierce arguments about how the precautionary principle should be applied in novel industrial processes (e.g. genetically modified crops, coal-seam gas extraction through 'fracking', nano technologies in food and cosmetics, and promotion of low-carbon technologies to mitigate changes caused by global warming). Moreover, the ongoing fundamental issues about the appropriate balance between economic development and ecosystem health tend to be inherently shaped by powerful underlying interests and ideologies. The role of scientific findings in such areas is heavily circumscribed. Science is better at calibrating the past than illuminating the future. According to Funtowicz and Ravetz (2003), major decisions about future risks need to be assessed and managed through much broader socio-political processes of education and deliberation.

Second, in the context of public policy, decisions by legislators are typically framed and decided through a mix of knowledge, interests, ideologies and cultural beliefs (Head 2008). Political leadership and judgements about political and fiscal feasibility are central to the policy-making process. The context of decision-making is inherently political. In addition, the practical knowledge and experience of organisational professionals (e.g. service delivery program managers, policy and regulatory executives) are vital for assessing administrative feasibility, evaluating past programs, engaging with stakeholders and reconciling conflicting priorities. Thus the contribution of science – understood as rigorous evidence about current programs and reliable data for analysing future policy options – is only one part of policy-relevant knowledge.

Third, the policy-making process does not generally centre on evidence as such (the knowledge content) but on how policy problems are framed or formulated, and thus how policy agendas are set (Rochefort and Cobb 1994; Fischer 2003; Boin *et al.* 2009). Science can inform the participants but cannot determine the outcome of such framing contests. Scientific evidence becomes a resource that can be selectively mobilised in support of various propositions. For example, when the Murray–Darling Basin in south-eastern Australia experienced a very prolonged drought, commencing around 1997, the irrigated agriculture industries lobbied to continue the existing high levels of extraction in order to maintain jobs and investment in the food production sector, whereas the environmental movement argued that the key priorities should be the ecological health of the waterways and the needs of downstream water users. CSIRO was commissioned by the Prime Minister to undertake an assessment of sustainable yields from the Murray–Darling river system; it was completed in 2008 (Hatton and Young 2011). The project primarily required biophysical modelling of water inflows/outflows over time (i.e. objectively calculating past experiences and estimating a range of future scenarios). But the subsequent policy responses could not have been deduced from the hydrological science. Policy-makers had to consider socio-economic impacts in local communities and the politics of how scarce water resources were to be apportioned between economic and environmental uses, under a range of water variability options. In practice, this became very contentious and required several further years of negotiation and socio-economic analysis (Connell and Grafton 2011). Importantly, the shaping of the final settlement in 2012 was fundamentally a product of complex socio-political bargaining and trade-offs rather than a product of the scientific models.

In short, the ways in which expert knowledge can influence the policy process are diverse and often indirect. While knowledge from the social sciences can be as valuable as findings from the biophysical sciences, the role of stakeholders and policy bureaucrats in determining policy and regulatory outcomes is very important. The competition of ideas is intense within policy debates and advisory systems (Craft and Howlett 2013), both in relation to defining and prioritising problems, and in relation to proposing programs of action (or inaction). There are perceived tensions between the expert contributions of scientific knowledge and the politicised clash of societal interests, which are managed through the process of democratic decision-making (Lee 1993). Ideally, some of the tensions can be diffused in the long run through adaptive management of programs and learning about what works.

While policy debates sometimes have the appearance of open discussion with many points of leverage, at other times the policy process is much more structured, with more focused consideration of key issues through expert submissions. The next section considers some of the roles available to science experts from across the disciplinary spectrum, in their quest to influence policy processes.

The specific role of scientific experts

There are many social science experts who may wish to influence the policy process directly, for example by communicating their policy-relevant scientific knowledge to practitioners involved in policy, regulation or service delivery. There are many ways to communicate expert knowledge and advice. A few prominent academics attain the status of 'public intellectual' and their views are widely sought and publicised, but most academics work through the quieter corridors of influence (Pollitt 2006; Jackson 2007). One avenue is contract research, where academic researchers may provide reports for government organisations seeking new information or new analysis on a specific issue. Other options arise in providing expert technical advice to government agencies about model-building, research methodology or quality assurance. At a broader level, academics may contribute to the gradual rethinking (or reframing) of an important public issue or socio-ecological challenge, making use of various channels such as their published writings, media communications and private conversations with key practitioners. Changing people's perspectives may take many years to become evident, but the long-term effects of paradigm change can be profound.

Most commonly, scientific advice is associated in the public mind with expert advisory bodies and processes (Stewart and Prasser 2015). There are numerous opportunities for social science experts to be involved in their work, either as members of expert panels or by making expert submissions to inquiries. The purpose, composition and longevity of such bodies is variable (OECD 2015; Wilsdon et al. 2014; Reillon 2015). Five types may be distinguished. The first type consists of the various research organisations and learned academies that are external to government; some of their members undertake contract research but the tone is primarily academic rather than instrumental research. Second, government agencies may establish specialised research units that employ skilled researchers in the social and biophysical sciences; these units often play important roles in translating both internal and external research findings as part of the briefing process for senior executives, ministers and legislators. Third, in some policy and regulatory areas, there are expert standing committees (with a majority of external members) that variously assess quality standards, risk levels and cost-effectiveness, and perhaps provide expert advice on emerging issues. Fourth, numerous *ad hoc* expert committees are formed to respond rapidly to contemporary crises or policy challenges; the most formal of these are public inquiries or royal commissions. Finally, legislative committees undertake a wide range of inquiries on specific topics relevant to human well-being and ecological health, and generally encourage specialist advice from across the sciences.

Some advisory bodies are mainly concerned with high-level strategic issues, such as the federal government's Commonwealth Science Council which has a strong focus on the STEM disciplines and their linkages to industry innovation systems (http://www.chiefscientist.gov.au/2014/11/commonwealth-science-council). By contrast, the Productivity Commission is a statutory body with substantial in-house resources with which to undertake evidence-intensive policy inquiries (http://www.pc.gov.au). Its processes usually include two rounds of expert submissions in response to issues papers and draft reports. Key topics have included environmental, economic and social issues, and the policy recommendations usually show a preference for market-based efficiencies. Given the heavily conflictual history of relations between economic development lobbies and the conservation movement, governments have occasionally established science-informed processes to assist in mediating between opposed positions and using best-available expertise from all relevant disciplines. The Resource Assessment Commission was established in 1989 to examine complex resource-use questions referred by the Prime Minister. In only four years it undertook significant public inquiries on major contentious issues including mining in Kakadu, forestry and timber resources, and sustainable management of the coastal zone, but its deployment of social and ecological science was seen as insufficient to resolve essentially political problems (Economou 1996). Other Australian government bodies, such as the Economic Planning Advisory Council (1983–96), have provided influential social science advice. Its secretariat played an influential role in social and economic policy development under a Labor government, before being merged with the Productivity Commission.

Regulatory agencies within government have a legal duty to use best-available information in undertaking work such as risk analysis, cost–benefit analysis and regulatory impact analysis. In the US, Jasanoff (1990) examined the challenges faced by authoritative bodies dependent on scientific advice, such as the Environmental Protection Agency and the Food and Drug Administration. She found that, in these regulatory settings, the distinction between pure science and policy advice may become somewhat artificial.

Social science expertise may be utilised in three main ways: to improve policy understanding and learning, to provide political justifications for preferred actions by government, and to provide specific technical instruments and solutions (Weiss 1979, 1999; Weible 2008). With rare exceptions, it is extremely difficult to measure the extent to which scientific advice and knowledge are adopted by decision-makers. In STEM disciplines where applied technologies are closely linked to commercial investment, the level of industry funding of the research and the successful commercialisation of intellectual property

Table 13.1. The use of expert-based information in three types of policy subsystem

	Unitary subsystems	Collaborative subsystems	Adversarial subsystems
Analytic compatibility	Experts agree on theory, data and methods	Experts reconcile differences in theory, data and methods	Experts disagree on theory, data and methods
Treatment of uncertainty and risk	Uncertainty used for political gains	Uncertainty acknowledged and decisions proceed adaptively	Uncertainty used for political gains
Experts and coalitions	Experts serve as auxiliary allies	Experts serve as auxiliary allies or opponents	Experts serve as principal allies or opponents
Policy-oriented learning	High intra-coalition learning and no inter-coalition learning	High intra-coalition learning and high inter-coalition learning	High intra-coalition learning and low inter-coalition learning

Source: Weible (2008, p. 628).

are convenient indicators of adoption. But the influence of the social sciences, which seldom attract commercial spin-offs, is generally more difficult to demonstrate, particularly in contexts dominated by technocentric worldviews. The provision of frameworks, evaluation criteria and diagnostic tests to guide analysis is a relatively common service to industry and community organisations. However, the provision of clear advice leading to policy solutions is quite a different type of policy-related work – and policy advocacy activities tend to attract disagreement among scientists. For example, Lackey (2007) drew a distinction between speaking as a scientist (objective findings) and speaking as an advocate (proposals for change), arguing that scientific credibility is inherent in the first role. Pielke (2007) distinguished between a range of detached and policy-engaged positions that might be adopted by scientists, and recommended the 'honest broker' role as most likely to preserve the desirable degree of credibility and salience. The politicised interpretation of scientific findings is no doubt inevitable, and beyond the control of researchers. Weible (2008) summarised three different institutional contexts in which scientific expertise is considered in policy deliberations (Table 13.1). The risk is that scientific work becomes ammunition in the adversarial policy wars.

The broad contributions of social science

The current and potential contributions of the social sciences for understanding and responding to sustainability issues are very substantial, but not as well appreciated as they deserve. The capacity of the social sciences to help in resolving very specific problems, or the capacity of researchers to gain direct influence on a hot issue with decision-makers, are underpinned by several more generic processes and capacities. Important challenges remain in better demonstrating these contributions, and improving decision-makers' understanding of how the social sciences add public value. This section briefly outlines six dimensions of social science work where increased public value can be generated and where priorities need to be focused. These may be summarised in terms of how social science contributes to:

- higher education and training for professionals and managers;
- productive interdisciplinary research, i.e. selectively harnessing social science expertise to work on sustainability challenges with the STEM disciplines;

- building research partnerships with government, industry and community organisations in order to enhance collaboration with potential end-users;
- expanding the role of knowledge-brokering and inclusive forums;
- raising the quality of decision-making and democratic debate by influencing the understanding of key participants and sections of the public;
- understanding the importance of appropriate governance arrangements for collaboration about problems and solutions.

The significant role of the social sciences in educating and training professionals and managers – who will be working in many sectors – should not be underestimated. Social science knowledge provides the fundamentals for understanding individual and organisational behaviour, and provides key methods for analysis and action. Engineers in the private sector who ignore the institutional contexts of funding, contracting, performance reporting, team culture and communication will not be commercially successful. Executive managers with social science expertise appreciate the importance of good processes such as transparency, accountability, consultation, knowledge sharing and building broad understanding among stakeholders.

Social science knowledge is also essential for addressing major socio-ecological problems, a perspective now widely acknowledged by leading experts from the STEM disciplines (Stern *et al.* 1992; Pohl 2008; Lang *et al.* 2012). This more integrated approach is becoming embedded in major interdisciplinary research centres, in major new funding programs and in numerous new journals of high quality such as *Ecology & Society* and *Global Environmental Change*. Recent work on how to increase research impact concluded that the key policy problems are inherently interdisciplinary and that the social sciences should adjust their priorities to play a full role in these evolving areas (Bastow *et al.* 2014). The history of power imbalances between the dominant STEM grouping and the social sciences needs to be recognised and adjusted if transdisciplinary approaches are to thrive (Holm *et al.* 2013).

The social sciences are taking positive steps to enhance their relationships with non-academic audiences. Social scientists have been building research partnerships with government, industry and community organisations in order to enhance collaboration with potential end-users (Hart *et al.* 2015). While these relationship-based initiatives have had only modest success in terms of research uptake (Head *et al.* 2014), much more can be done to build bridges, produce joint agendas for knowledge production in priority areas, and develop new actionable knowledge.

Knowledge brokering is vital for developing shared focus, understanding and action. Knowledge brokering comprises various mechanisms for sharing knowledge and enhancing understanding across the boundaries of organisations, professions and disciplines (Cash *et al.* 2003; Davies *et al.* 2008). The approaches selected need to be adapted for the scale of the issue, the organisational contexts and the stakeholders (Michaels 2009). The practices of knowledge brokering go beyond the old 'transmit and disseminate' models of science communication (where researchers inform others about new findings). Knowledge brokering seeks to add value for end-users through tailored dialogue and co-production of insights relevant to particular challenges of mutual interest. While much knowledge-brokering activity is directed towards decision-makers and professionals, there are many advantages in drawing a wider net of stakeholders into informed dialogue about sustainability issues. For example, one instructive experiment has been the Healthy Waterways Report Card, developed over many years to inform the community about water quality in coastal streams and marine estuaries in south-east Queensland. Initially the scorecard focused on biophysical aspects of water quality, but the most recent version includes feedback about the socio-economic benefits of waterways and how those benefits would be affected by changing environmental conditions:

> *The following components are measured: community values and satisfaction with waterways; appropriate access to local waterways; economic benefits generated through recreation; economic benefits provided through recreational fishing; clean waterways that support the supply of low cost drinking water (http://healthywaterways.org/reportcard/#/overview/benefits).*

The social sciences play a large role in raising the quality of decision-making and democratic debate, by providing knowledge and insights which influence the understanding of key participants and sections of the public. Contributing to the quality of information and debate is central in democratic societies, and harks back to the enlightenment function of social science envisaged by foundational thinkers such as John Dewey in the 1920s and Harold Lasswell in the 1950s. One of the invaluable roles of social science is in building stakeholder understanding and capability, and in this way contributing to improved design and implementation of well-crafted initiatives. Moreover, the participatory methods used by many social scientists may help to increase the salience and legitimacy of their findings and their ultimate influence in decision-making processes.

Finally, the social sciences provide essential frameworks for understanding the design of appropriate governance arrangements for understanding and addressing important problems, effectively implementing agreed programs, and testing the efficacy of solutions (Adger and Jordan 2009; Cash *et al.* 2003; Kenny and Meadowcroft 1999; Stern *et al.* 1992). Social science insights are essential for making evidence-informed policy choices, increasing community resilience in the face of disruptive change, and building adaptive capacity in all sectors (Dovers and Wild River 2003; Barnett *et al.* 2011).

Conclusion: engagement and communication

The field of sustainability science posits an active orientation. Most of the literature agrees with the view that:

> *sustainability science must link research on problem structures with a solutions-oriented approach that seeks to understand, conceptualize and foster experiments for how socio-technical innovations for sustainability develop, diffuse and scale up (Miller et al. 2014, p. 240).*

This amounts to a focus on developing policy and socio-technical innovations that work to address real problems. The social, economic and institutional dimensions of change (e.g. for adapting to the impacts of climate change) need to be well managed on the basis of best-available information and shared perspectives. Some issues, such as point-source pollution, may be best tackled at a local scale, while others such as water basin management may need to be analysed and addressed at a regional or national scale (Head *et al.* 2016). Some issues benefit from global frameworks concerning strategic goals and programs for collective action, such as the UN's eight Millennium Development Goals for 2000–15, and the even more ambitious 17 Sustainable Development Goals for the years to 2030 (http://www.undp.org/content/undp/en/home/mdgoverview.html).

Researchers need to consider how their contributions can produce credible, salient and legitimate science (Cash *et al.* 2003), that is respected by stakeholders and decision-makers. Wittmayer and Schäpke (2014) suggested that a concern for societal learning should be central in sustainability research, and that careful choices need to be made about how to position the roles of researchers as reflective scientists, change agents or facilitators of deliberative processes. As the social sciences attempt to improve their capacity to provide trusted advice and thereby gain policy influence, key requirements will include the capacity to provide high-quality analysis and build collaborations with potential end-users. By focusing on important challenges where improved knowledge can contribute to problem-solving, the invaluable role of social science in improving implementation success can be demonstrated.

References

Adger WN, Jordan A (Eds) (2009) *Governing Sustainability*. Cambridge University Press, Cambridge.

Barnett J, Dovers S, Hatfield-Dodds S, McDonald J, Nelson R, Waller S (2011) *National Climate Change Adaptation Research Plan: Social, Economic and Institutional Dimensions*. National Climate Change Adaptation Research Facility, Gold Coast. <https://www.nccarf.edu.au/publications/national-climate-change-adaptation-research-plan-social-economic-and-institutional>.

Bastow S, Dunleavy P, Tinkler J (2014) *The Impact of the Social Sciences: How Academics and Their Research Make a Difference*. Sage, London.

Boin A, 't Hart P, McConnell A (2009) Crisis exploitation: political and policy impacts of framing contests. *Journal of European Public Policy* **16**(1), 81–106. doi:10.1080/13501760802453221

Cash DW, Clark W, Alcock F, Dickson N, Eckley N, Guston D, Jäger J, Mitchell R (2003) Knowledge systems for sustainable development. *Proceedings of the National Academy of Sciences of the United States of America* **100**(14), 8086–8091. doi:10.1073/pnas.1231332100

Connell D, Grafton RQ (2011) Water reform in the Murray-Darling Basin. *Water Resources Research* **47**(12), 1–9. doi:10.1029/2010WR009820

Craft J, Howlett M (2013) The dual dynamics of policy advisory systems: the impact of externalization and politicization on policy advice. *Policy and Society* **32**(3), 187–197. doi:10.1016/j.polsoc.2013.07.001

Davies HT, Nutley S, Walter I (2008) Why 'knowledge transfer' is misconceived for applied social research. *Journal of Health Services Research and Policy* **13**(3), 188–190. doi:10.1258/jhsrp.2008.008055

Dovers S, Wild River S (Eds) (2003) *Managing Australia's Environment*. Federation Press, Sydney.

Economou N (1996) Australian environmental policy making in transition: the rise and fall of the Resource Assessment Commission. *Australian Journal of Public Administration* **55**(1), 12–22. doi:10.1111/j.1467-8500.1996.tb01178.x

Fischer F (2003) *Reframing Public Policy: Discursive Politics and Deliberative Practices*. Oxford University Press, New York.

Funtowicz S, Ravetz J (2003) *Post-normal Science*. Internet Encyclopaedia of Ecological Economics. <http://isecoeco.org/pdf/pstnormsc.pdf>.

Hart D, Bell K, Lindenfeld L, Jain S, Johnson T, Ranco D, McGill B (2015) Strengthening the role of universities in addressing sustainability challenges: the Mitchell Center for Sustainability Solutions as an institutional experiment. *Ecology and Society* **20**(2), 4. doi:10.5751/ES-07283-200204

Hatton T, Young W (2011) Delivering science into public policy: an analysis of Murray-Darling Basin sustainable yields assessment as a model for impact. *Australian Journal of Public Administration* **70**(3), 298–308. doi:10.1111/j.1467-8500.2011.00730.x

Head BW (2008) Three lenses of evidence-based policy. *Australian Journal of Public Administration* **67**(1), 1–11. doi:10.1111/j.1467-8500.2007.00564.x

Head BW (2014a) Managing urban water crises: adaptive policy responses to drought and flood in south-east Queensland. *Ecology and Society* **19**(2), 33. doi:10.5751/ES-06414-190233

Head BW (2014b) Evidence, uncertainty and wicked problems in climate change decision-making in Australia. *Environment and Planning. C, Government and Policy* **32**(4), 663–679. doi:10.1068/c1240

Head BW (2015) Policy analysis: evidence based policy-making. In *International Encyclopedia of the Social and Behavioral Sciences*. (Ed. JD Wright) 2nd edn, Vol. 18, pp. 281–287. Elsevier, Oxford.

Head BW, Alford J (2015) Wicked problems: implications for public policy and management. *Administration and Society* **47**(6), 711–739. doi:10.1177/0095399713481601

Head BW, Ferguson M, Cherney A, Boreham P (2014) Are policy-makers interested in social research? Exploring the sources and uses of valued information among public servants in Australia. *Policy and Society* **33**(2), 89–101. doi:10.1016/j.polsoc.2014.04.004

Head BW, Ross H, Bellamy J (2016) Managing wicked natural resource problems: the collaborative challenge at regional scales in Australia. *Landscape and Urban Planning* **154**, 81–92. doi: 10.1016/j.landurbplan.2016.03.019

Holm P, Goodsite M, Cloetingh S *et al.* (2013) Collaboration between the natural, social and human sciences in global change research. *Environmental Science and Policy* **28**, 25–35. doi:10.1016/j.envsci.2012.11.010

Jackson PM (2007) Making sense of policy advice. *Public Money and Management* **27**(4), 257–264. doi:10.1111/j.1467-9302.2007.00592.x

Jasanoff S (1990) *The Fifth Branch: Science Advisers as Policymakers*. Harvard University Press, Cambridge, MA.

Kates RW (2011) What kind of a science is sustainability science? *Proceedings of the National Academy of Sciences of the United States of America* **108**(49), 19449–19450. doi:10.1073/pnas.1116097108

Kates RW, Clark W, Corell R *et al.* (2001) Sustainability science. *Science* **292**(5517), 641–642. doi:10.1126/science.1059386

Kenny M, Meadowcroft J (Eds) (1999) *Planning Sustainability*. Routledge, London.

Lackey RT (2007) Science, scientists, and policy advocacy. *Conservation Biology* **21**(1), 12–17. doi:10.1111/j.1523-1739.2006.00639.x

Lang D, Wiek A, Bergmann M, Stauffacher M, Martens P, Moll P, Swilling M, Thomas C (2012) Transdisciplinary research in sustainability science: practice, principles, and challenges. *Sustainability Science* **7**(Supp. 1), 25–43. doi:10.1007/s11625-011-0149-x

Lee KN (1993) *Compass and Gyroscope: Integrating Science and Politics for the Environment*. Island Press, Washington DC.

Lindblom CE, Cohen D (1979) *Usable Knowledge: Social Science and Social Problem Solving*. Yale University Press, New Haven.

Michaels S (2009) Matching knowledge brokering strategies to environmental policy problems and settings. *Environmental Science and Policy* **12**(7), 994–1011. doi:10.1016/j.envsci.2009.05.002

Miller T, Wiek A, Sarewitz D, Robinson J, Olsson L, Kriebel D, Loorbach D (2014) The future of sustainability science: a solutions-oriented agenda. *Sustainability Science* **9**(2), 239–246. doi:10.1007/s11625-013-0224-6

OECD (2015) *Scientific Advice for Policy Making: The Role and Responsibility of Expert Bodies and Individual Scientists*. OECD, Paris.

Oreskes N (2004) Science and public policy: what's proof got to do with it? *Environmental Science and Policy* **7**(5), 369–383. doi:10.1016/j.envsci.2004.06.002

Palmer MA (2012) Socioenvironmental sustainability and actionable science. *Bioscience* **62**(1), 5–6. doi:10.1525/bio.2012.62.1.2

Pielke RA Jr (2007) *The Honest Broker: Making Sense of Science in Policy and Politics*. Cambridge University Press, Cambridge.

Pohl C (2008) From science to policy through transdisciplinary research. *Environmental Science and Policy* **11**(1), 46–53. doi:10.1016/j.envsci.2007.06.001

Pollitt C (2006) Academic advice to practitioners: what is its nature, place and value within academia? *Public Money and Management* **26**(4), 257–264. doi:10.1111/j.1467-9302.2006.00534.x

Reillon V (2015) *Scientific Advice for Policy-makers in the European Union*. Briefing paper. European Parliamentary Research Service.

Rochefort D, Cobb R (Eds) (1994) *The Politics of Problem Definition*. University of Kansas Press, Lawrence.

Sayer A (2011) *Why Things Matter to People: Social Science, Values and Ethical Life*. Cambridge University Press, Cambridge.

Stern P, Young O, Druckman D (Eds) (1992) *Global Environmental Change: Understanding the Human Dimension*. National Academy Press, Washington DC.

Stewart J, Prasser S (2015) Expert policy advisory bodies. In *Policy Analysis in Australia*. (Eds BW Head, K Crowley) pp. 151–166. Policy Press, Bristol.

Weible CM (2008) Expert-based information and policy subsystems: a review and synthesis. *Policy Studies Journal: The Journal of the Policy Studies Organization* **36**(4), 615–635. doi:10.1111/j.1541-0072.2008.00287.x

Weiss CH (1979) The many meanings of research utilization. *Public Administration Review* **39**(5), 426–431. doi:10.2307/3109916

Weiss CH (1999) The interface between evaluation and policy. *Evaluation* **5**(4), 468–486. doi:10.1177/135638909900500408

Weiss CH, Bucuvalas M (1980) Truth tests and utility tests: decision-makers' frames of reference for social science research. *American Sociological Review* **45**(2), 302–313. doi:10.2307/2095127

Wilsdon J, Allen K, Paulavets K (2014) *Science Advice to Governments: Diverse Systems, Common Challenges*. Briefing paper for Auckland conference on Science Advice.

Wittmayer JM, Schäpke N (2014) Action, research and participation: roles of researchers in sustainability transitions. *Sustainability Science* **9**(4), 483–496. doi:10.1007/s11625-014-0258-4

Index

Aboriginal Australians *see* Indigenous Australians
abstraction of expectations 17
accident avoidance 192
accidents, catastrophic industrial 190–4
adaptability 54–5, 63
adaptation pathways 52, 63, 116, 121
adaptation practices 119–20
Adapting to Climate Change in Asia (ACCA) 119–20
adaptive capacity, 54–5, 59, 60, 62, 63, 101–2, 104, 105
   of rural communities 121
   Sydney 99–106
   and vulnerability 101–2, 104–5
adaptive learning 122–3, 124
Adelaide, stormwater recycling 150
   water supply 146–7
advisory bodies 202, 203
agencies, regulatory 203
agriculture, nutrition-sensitive 113, 181
Aichi Targets 26, 35
Alice Springs 77–9, 87
Alice Springs Desert Park 87–8, 89, 90
Anmatjere region, employment 85–7, 88, 89
Applied Research and Innovation Systems in Agriculture (ARISA) 114
Atomic Energy Commission (India) 193
Atomic Energy Regulatory Board (AERB) 193
AusAID 113
Australian Centre for International Agricultural Research (ACIAR) 113, 118, 181
Australian Indigenous biocultural knowledge (IBK) 27, 38
Australian International Food Security Research Centre (AIFSRC) 181

Bangladesh 118, 119
base flow 134
behaviour, after disasters 105
   changes in 40
   and climate change opinion 170–3
   and views of others 168–70
Beijing Platform for Action 179
beliefs, changes in 39
Bhopal disaster 190, 194
Biocultural Mapping for Climate Change 30, 31, 34, 36
boundary objects 182
boundary partners 121
bridging institutions 79, 85, 89–91
brokering, 39–40, 113–14, 178, 182–5, 205
   and gender 183–5
brokers 88–91
bush food, trade 83–4, 88, 89, 90
bush schools 84–5, 88–9, 90
bush tomato 82

Cambodia 119
capability 80, 81, 89
capacities, cognitive domain 35–6, 43
   co-productive 25, 30, 36, 42
   material domain 36, 38, 42, 43
   normative domain 31, 34, 42
   social domain 34–5, 42
capacity, adaptive 54–5, 59, 60, 62, 63, 101–2, 104, 105
capital, access to 53
   cultural 62
   and Indigenous livelihood activities 82–3
   political 53, 62, 64
   types 82–3
carbon abatement 40
Carbon Farming Initiative Indigenous Leaders' Roundtable 34

carbon tax   171, 172
catastrophic system failure   190
change, incremental   120–2
    transformational   24, 26, 179
    transformative   120–2
Chilean earthquake 2011   105
Clean Energy Legislative Package
    (CELP)   171, 172
Climate Adaptation Flagship (CSIRO)
    161
climate change, Australians' responses
    161–74
    and public opinion   161–74
    and voting behaviour   171–2
climate change mitigation   40
climate change vulnerability   102
climate-related behaviour, and
    opinions   170–3
coastal inundation   104
cognitive dissonance theory   167, 168, 170
cognitive domain capacities   35–6, 43
Colorado River Basin   133–6, 138, 139, 140
commercial harvesting, Indigenous   82–4,
    88, 89, 90
commonality   30–1
Commonwealth Science Council   203
communication, of risk   193–4
    social   14–16
    and water use   157
community dialogues   35–6
community preferences, and water
    supply   146–7
community responses, to water reuse
    147–50
complexity   14, 18
content analysis   134–5, 139
contextual knowledge   118
co-productive capacities   25, 30, 36, 42
cross-scale linkages   23, 36, 38, 103
CSIRO   2–3, 110, 113, 114, 118, 124, 161, 183,
    184, 201
cultural capital   62
cultural ecosystem benefits   132, 135–40
cultural ecosystem services   132, 133,
    138–40
cultural evolution   12, 14, 20
cultural theory   173
culture, workplace   85–8

decentralised water systems   150–2
decision-making, policy   200–2
Deepwater Horizon accident   194
denial, of climate change   168
Department of Atomic Energy (DAE)
    (India)   193
Department of Foreign Affairs and Trade
    (DFAT)   113, 114, 119, 123, 181, 182
desert raisins, trade   83–4, 88, 89, 90
development problems   109–10
development, research for   109–24
dialogues, community   34, 35–6, 114, 149,
    182, 185, 205
disasters, community behaviour   105
    and social relations   105–6
dissociation   168
distrust, public   193
double-loop learning   116, 117, 121

East Gangetic Plains, food systems   118–19
ecosystem services   51, 55, 60, 62, 132, 133,
    138–40
education, Indigenous   84–5, 88–9, 90
elderly, vulnerability   100–1
employment, Indigenous   81–8, 89, 90
enlightenment   20
environmental flow   134, 139
environmental problems, and
    sociology   9–12
environmental risk   104, 173
ethics, changes in   38
evaluation, of R4D   122–3
expert knowledge, and policy-making
    200–4
experts, and policy-making   202–4
exposure, to vulnerability   101–2
extreme heat events, vulnerability   100–1

false consensus   168
financial capital   83
food and nutrition security   178, 179, 181,
    182
food security   55, 56, 113, 181
food systems, East Gangetic Plains   118–19
    research   118–20
Food Systems Innovation (FSI)   113, 114,
    122–3, 180–2, 183
Frankfurt School   11, 13, 20

Fukushima Daiichi disaster 187–8, 191, 193, 195
functional differentiation 13–18

gender, and brokering 183–5
  and nutrition 181–2
  and research citation 180
  and sustainability science 177
  and vulnerability 101
gender equality 179–80
gender gaps 180
gender mainstreaming 179–80
gender parity 180
Global Alliance for Improved Nutrition (GAIN) 178
governance, of risk 193, 195
government policy, support of 171–2
Great Barrier Reef 27, 30, 31, 36, 39
grounded theory 135
Guna people 31, 35

Habermas, Jürgen 9, 10, 11–14, 19, 20
harvesting, Indigenous 82–4, 88, 89, 90
Healthy Waterways Report Card 205–6
hierarchy of needs, for development 109–24
High Reliability Organization (HRO) research 192
homeless, vulnerability of 101
household water demand, Queensland 153–4
human capital 82–3
Hurricane Katrina 189

implementation science 178–9
inclusiveness 179
income disparity 105
incremental change 120–2
India 118, 119, 120, 190, 193, 194
Indigenous Advisory Committee to the Minister for Environment 36
Indigenous and local knowledge (ILK) 26, 31
Indigenous Australians, demography 76–7
Indigenous communities, remoteness 76–9
Indigenous cultures, desert 75–92
Indigenous ecosystem services 132
Indigenous education 84–5, 88–9, 90
Indigenous employment 81–8, 89, 90

Indigenous inequality 80
Indigenous knowledge 23, 26, 27, 30, 31, 36, 38, 39, 89
Indigenous land management 40, 42, 89, 91
Indigenous land ownership 39
  see also Indigenous Traditional Owners
Indigenous Language and Culture (ILC) programs 84–5, 89, 90
Indigenous livelihood activities, and capitals 82–3
Indigenous livelihoods 75–92
Indigenous mobility 78–9
Indigenous people, and carbon abatement 40
  marginalisation 23, 40, 42, 80
Indigenous plant foods, harvesting 82–4, 88, 89, 90
Indigenous Protected Area program 90
Indigenous recognition 39, 42
Indigenous research 42
Indigenous rights 39, 40, 42
Indigenous social disadvantage 75, 76–7
Indigenous teaching assistants 85, 89
Indigenous Traditional Owners 30, 31, 34–5, 36, 40, 87–8
Indigenous workplace culture 85–8
industrial accidents, catastrophic 190–4
industrial pollution 190
inequality, Indigenous 80
innovation, levels 119–20
innovation systems 178, 180
Institute for Social-Ecological Research 11, 13
institutional analysis 61–2, 36
institutions, bridging 79, 85, 89–91
  changes in 39–40
  recognition of 51
  traditional 55, 56, 59, 60
integration science 23–43
intellectual capital 82–3
Intergovernmental Platform on Biodiversity and Ecosystem Services (IPBES) 26, 39
intervention, levels 119–20
interviews 102, 103, 134, 151, 153
inundation risk 104
Isan people, Thailand 59–62, 71–3

knowledge, contextual 118

Indigenous   23, 26, 27, 30, 31, 36, 38, 39, 89
  working   31
knowledge brokering   205
knowledge integration   31, 34, 65, 113, 117–19, 178, 180
knowledge types   30, 117–18
Kowanyama Aboriginal Land and Natural Resource Management Office   30

land management, Indigenous   40
land ownership, Indigenous   39
  see also Indigenous Traditional Owners
Laos   119, 120
learning, adaptive   122–3, 124
  double-loop   116, 117, 121
  multi-loop   116
  and R4D   122–3
  social   115, 118, 121, 122, 180
  triple-loop   116, 117, 121
Leveraging Agriculture for Nutrition in South Asia (LANSA)   178
lifeworld   12, 13, 19
linkages, cross-scale   23, 36, 38, 103
livelihood adaptation strategies   63–4
livelihood analysis, and social-ecological systems (SES)   42–52, 62–4
livelihood systems, and resilience   79–81
  and sustainability   79–88
livelihoods, Indigenous   75–92
  sustainable   53, 56
  and sustainable livelihoods framework (SLF),   62–4
local government, and vulnerability   103, 104–5
Lower Hunter Water Plan   157
Luhmann, Niklas   9, 10, 13–17, 19, 20

Macondo oil rig   191
maintenance, of rainwater tanks   151–2
malnutrition   113, 181
managed aquifer recharge   149–50
marginalisation, of Indigenous people   80
Marx, Karl   11, 13
material domain capacities   36, 38, 42, 43
migrant labour   55, 56, 60
migration,   56, 59, 60, 61, 72, 73, 118
  in remote Australia   78–9

Millennium Development Goals   1, 110–11, 206
Millennium drought   145, 150, 151, 152
Minute 319 pulse flow   133, 134
mobility, Indigenous   78–9
modelling, participatory   35, 43
modernisation   91
Mongolia   113
monitoring, evaluation and learning (MEL) of R4D   122–3, 124
multi-loop learning   116
Murray–Darling Basin   146, 147, 201–2

NASA   191
National Indigenous Climate Change Project (NICC)   27, 34, 40
National Research Council (USA)   200
National Reserve System   39
native title   39, 87
natural capital   62, 63, 83
natural disasters   101, 105
nature–society interrelationships, research   134–8
Nepal   118
New Orleans   189–90
normative domain capacities   31, 34, 42
North Sea gas and oil platform explosion 1988   191–2
Northern Territory Curriculum Framework (NTCF)   84
NTB Climate Futures (Indonesia)   116, 121–2
nuclear industry (India)   193
nutrition security   178, 179, 181, 182
nutrition-sensitive agriculture   113, 181

Official Development Assistance (ODA) program   110, 113, 114, 124
Ogiek people   39
oil spills   194
One Health   178
ontologies   30
opinion (public), of climate change   161–74
opinions, and climate-related behaviour   170–3
Oriners Station   30, 31, 35
Ostrom, Elinor   82, 86

participation by stakeholders, in research   115–17
participatory modelling   35
partnership agreements   114–15
partnerships, in research   114–15
  transactional   115
  transformational   115
path-generation, transformative   23, 24, 26, 38–40
payment for ecosystem service (PES) schemes   27, 40, 132
Perth, wastewater use   149–50
physical capital   82, 83
place,   99
  linkages across   104–5
place-based connections   31, 104
pluralistic ignorance effect   168
policy-making process,   200–2
  and expert knowledge   200–4
political capital   53, 62, 64
pollination assessment   31, 34, 38, 39
poverty   54, 55, 56, 62, 63, 75, 80
poverty traps   52, 121
Productivity Commission   203
public opinion, of climate change   161–74
public trust, and risk   193–4
pulse flow   133–40
purified recycled water (PRW)   148, 149, 150

QMethodology   149
Queensland, wastewater recycling   147–9

R4D   109–24, 178, 179, 180
rainwater tanks   146, 150–2, 153
recognition of institutions   51
recycling, stormwater   146, 150
  wastewater   147–50, 155
region   99
regulatory agencies   203
regulatory capture   193
remoteness, Indigenous communities   76–9
research, food systems   118–20
  Indigenous   42
  policy-relevant   199–207
  stakeholder participation   115–17
research for development (R4D)   109–24, 178, 179, 180

research partnerships   114–15
research skills   111, 113–14, 124
resilience, in livelihood systems   79–81
resilience-thinking,   51, 62, 64, 65
  and social-ecological systems (SES)   54–5, 56, 59–60
resonance   19, 20
Resource Assessment Commission   203
restitution, Indigenous land   39
restoration flow   133–40
risk, coastal inundation   104
  distribution   189–90
  environmental   104, 173
  governance   193, 195
  inequalities of   189–90
  and public trust   193–4
  and regulatory institutions   192–3
  social amplification   194
  and social assumptions   189
risk assessment, socially informed   194–6
risk calculation, technical   188–9
risk communication   193–4
risk escalator   194–5
risk governance framework   195–6
risk management,   187–8
  socially informed   194–6
rivers, relations with human society   131
rural communities, adaptive capacity   121

safety, and regulatory institutions   192–3
safety culture   192
Sang Saeng village, Thailand   55–6, 59–62, 71–3
schools, bush   84–5, 88–9, 90
scientific experts, and policy-making   202–4
scientific knowledge, and policy-making   201–4
sensitivity, to vulnerability   101–2
Sieferle, Rolf Peter   10
social amplification of risk   194
Social and Economic Long-term Monitoring Program (SELTMP)   27, 30, 31, 35, 39, 40
social assumptions, and risk   189
social capital   56, 63, 80–1, 83, 84, 86, 100, 105, 136, 138, 157
social change   16–17, 18

social disadvantage, Indigenous   75, 76–7
social domain capacities   34–5, 42
social learning   115, 118, 121, 122, 180
social relations, after disasters   105–6
social relationships with water   131–3
Social Risk Amplification Framework (SARF)   193–4
social sciences, research partnerships   205
social vulnerability   101, 102
social-ecological systems (SES),   9, 11, 26, 51–2, 55–6, 60–5
 and livelihood analysis   42–52, 62–4
 and resilience-thinking   54–5, 56, 59–60
social-ecological theory   10, 11
socially informed risk assessment   194–6
socially informed risk management   194–6
society-nature system   2, 9, 13, 15, 29
socio-cultural life   11–13
sociological system theory   13–18
sociology, and environmental problems   9–12
 and sustainable development   9–20
socio-technical systems   12, 18, 206
Solanum centrale   83, 84
Space Shuttle Challenger disaster   191
stakeholder participation, in research   115–17
STEPS Centre   36, 38
stormwater   147, 152, 155
stormwater recycling   146, 150
structural coupling   14–15, 18
sustainability, and livelihood systems   79–88
sustainability science,   2, 3–6, 23, 103, 105–6, 177–8, 184, 185, 199, 206
 and gender   177
 in the spatial context   99–101
Sustainable and Resilient Farming System Intensification (SRFSI)   118–19
Sustainable Development Goals   1, 111, 206
sustainable development, and sociology   9–20
sustainable livelihoods   53, 56
sustainable livelihoods framework (SLF)   51, 52, 55–6, 60–5, 75, 79, 81–2
Sydney, adaptive capacity   99–106

 vulnerability in   99–106
Sydney Coastal Councils Group (SCCG)   102, 104

tanks, rainwater   146, 150–2, 153
teaching assistants, Indigenous   85, 89
technical risk calculation   188–9
Theories of Change and Impact Pathways   123
Timor-Leste   179
trade, of desert raisins   83–4, 88, 89, 90
traditional institutions   55, 56, 59, 60
Traditional Owners, Indigenous   30, 31, 34–5, 36, 40, 87–8
transactional partnerships   115
transboundary flow   133, 139
transformational change   24, 26, 179
transformational partnerships   115
transformative change   53, 120–2
transformative path-generation   23, 24, 26, 38–40
triple-loop learning   116, 117, 121
trust, after disasters   105
 and water supply   156–7

UN Declaration on the Rights of Indigenous Peoples   39
Union Carbide India   190
urban design, water-sensitive   152–3
urban water security   145–57
urban water supply   145–57
urbanisation   1

values, changes in   39
Vienna School of Social Ecology   10, 13
voting behaviour, and climate change   171–2
vulnerability,   54, 62, 63, 99–106, 121
 and adaptive capacity   101–2, 104–5
 exposure to   101–2
 extreme heat events   100–1
 and gender   101
 sensitivity to   101–2
 social   101, 102
vulnerability mapping   102

wastewater   145–50, 152, 155
wastewater recycling   147–50, 155

water, and society   131–3
water demand, management   153–5
water entitlements, Colorado River   134
water resources, management   131–40
water reuse, and community
    responses   147–50
water security, technological solutions   145
    urban   145–57
water supply, Adelaide   146–7
    and community preferences   146–7
    and community trust   156–7
    urban   146–57
water systems, decentralised   150–2
water tanks   146, 150–2, 153
water use, and communication to users   157
    household   153–5

waterscapes   131
water-sensitive urban design (WSUD)   152–3, 155, 157
wealth disparity   56
Western Australia Water Corporation   156
Wet Tropics NRM cluster climate change
    adaptation project   30, 31, 36, 39
WNB Climate Futures (Papua New
    Guinea)   116
working knowledge   31
Working Knowledge project   30, 31
Working on Country program   90, 91
workplace culture   85–8
workshops   26–7, 31, 34, 35–6, 102, 116, 117, 121, 122, 157
worldviews   162, 173, 204

www.ingramcontent.com/pod-product-compliance
Lightning Source LLC
Chambersburg PA
CBHW040003040426
42337CB00033B/5213